GREENING STEAM

How to Bring 19th-Century Heating Systems
into the 21st Century *(and save lots of green!)*

Dan Holohan

Visit www.HeatingHelp.com for additional copies.

For Erin Holohan Haskell,
who is the music to my words.

CONTENTS

CHAPTER 1

GREENING STEAM

So here's a good question for us to consider; and I think this is a fine place for me to ask it – right up front. Let's get it out of the way now, and maybe I'll ask it again when we're all done chatting. See how we feel about it then.

Sound good? Okay, here's the question:

Why the heck, in the 21st Century, would anyone want to heat a building with 19th-Century technology?

Good question, right? There's so much modern heating equipment available these days, and all of the manufacturers of these products say their stuff will save fuel and be good for the planet, but what manufacturer doesn't say that nowadays?

Speaking of which, how about that term "green"? Manufacturers are using that a lot these days. It sounds so nice and fresh, and so friendly. But what the heck does green mean?

That's another good question to ask right up front (especially since it's part of the title of this book). What makes a heating system green? And green compared to what? Can a 19th-Century heating system be green? And *how* green should it be?

There are so many steam systems in service, and I think they'll be around for some years to come. It's not that easy (or inexpensive) to rip out a steam system and start anew, and that's why so many of them are still around. So how about if we look at ways to make them better?

And the good news is that most of these ways won't cost much green.

Here's what I mean.

Dead Men

In 1970, when I was 20 years old, I started my career in the heating industry by taking a clerk's job at the company where my father worked. He had said to me, "Kid, this is the best business to be in because people are always going to need heat, especially in the winter." That made sense to me then, and it still does now. People are *always* going to need heat. Especially in the winter.

We were a manufacturers' representative and my job at first was to check shipments of copper fittings, sent from the Northern Indiana Brass Company in Elkhart, Indiana to wholesalers in the New York/Metropolitan area. We didn't stock those fittings, and I spent my days looking at numbers on paper. It wasn't very exciting, but my father sat next to me in a tiny office, and he was always a grand Irish storyteller. He had come up through the supply-house side of the business, and during the 10 years we worked together at that company, he told me stories about the people he had known in the business. He told me about how, after World War II, when America was once again building, a heating contractor could walk into a supply house and drop off the plans for a building on the heating-guy's desk. "By the end of the day," my father told me, "the heating guy would have figured out and sized everything the contractor needed. And he would explain in plain English how to put it all in. Those guys knew *all* the tricks."

I wanted to know what they knew. I started to question the old heating guys as I met them, and I began to go to the big public library on Fifth Avenue and 42nd Street in New York City to read old books about heating. And that's when I met the Dead Men.

Those long-gone authors told me about a time when people were cold (*especially* in the winter). They told me how they had tinkered and invented and patented these wonderful steam-heating devices. The Dead Men were alive to me, and I met more and more of them as the years went by. Each had a story to tell. The more I read, the more I learned about what happened to those steam-heating devices as time passed. I learned what worked and what didn't work. I learned about what the public accepted and what they rejected. I felt like I had a time machine because I knew where all of this technology was going. I knew.

I worked at that company for 19 years, and spent a lot of time in the field, working with contractors. It was my job to figure out what that strange object in that dark basement was – that big, cast-iron device with the 19th-Century patent date on its side. You see we sold steam- and hot-water heating equipment. We had ways to make the old stuff work better. My job was to figure out how to do it on each job because each job was different. I learned that, most of the time, all it took was a bit of tweaking. The goal was to undo what so many people who didn't know what they were doing had done over the years. The goal was to put it back to the way it was supposed to be – to de-knucklehead it. Once I had figured that out, the simplicity of it all was positively delightful. Often, it was as simple as a clogged air vent.

I once asked my old boss where all those one-pipe-steam air vents were going. No one had installed a new one-pipe steam system since, what, 1930 or so? And there we were, all these years later, shipping truckloads of air vents to heating wholesalers all over New York City. "Is there a landfill in Brooklyn where they're putting all these air vents?" I asked. And my boss smiled and said, "Dan, as long as there are painters, we will sell one-pipe steam air vents."

And even though I've been gone from that company for more than 20 years, they still sell air vents. Those painters are relentless, and when they paint over the hole in the air vent, the air stops venting, and when the air stops venting, the radiator stops heating.

And often, the solution involves nothing more than a paperclip. Read on:

Doctor Bob's radiator

Our family dentist, Doctor Bob, bought a retirement home out on the east end of Long Island a few years back and he'd tell me about it while he had his hands in my mouth. "It's got steam heat," he'd say. "I was thinking of you when we bought it."

"Ath nyth," I'd answer, through his fingers.

"There's just one problem, though. The radiator in the living room doesn't heat well."

"Ahh uhh," I responded.

"Any idea what might be wrong with it?"

"Obale a ad aar ent." I explained.

"Probably a bad air vent, eh? I see," he said. "Can I get those at The Home Depot?"

"Abee, abee ot." I choked.

"I see," he said. "Well, do you want to come and visit?"

And since it's my policy to never say no to Doctor Bob while he has metal instruments in my mouth, I took a nice drive out to the end of Long Island.

Doctor Bob's house is a just block from the center of this tiny town that Norman Rockwell could have painted. The house has a one-pipe-steam system and an enormous cast-iron radiator in the living room. Doctor Bob was looking at me and I was looking at that huge hunk of iron in total appreciation for the lost art that is steam heating.

"How do you repaint something like this?" he said.

"Do you want to take the old paint off first?"

"Yes," he said.

"Hire someone to remove it and take it to a sandblaster," I said. "When they're done, have it powder coated. It will be gorgeous."

"Sounds good," he said.

"Bring money," I added.

"Will do," he said. "And expect your dental prophylaxis bill to increase significantly in the coming months."

He looked at the side of the radiator. "I don't see an air vent," he said. "That's what's confusing me. You said it might be a bad air vent but this radiator isn't like the others. This one has no vent."

I looked a little closer and he was right, but that didn't make sense because you can't have a one-pipe-steam radiator without an air vent. Well, that's not exactly true; you *can* have a one-pipe steam radiator without an air vent. It just won't work.

I looked closer, and then I saw it. It was there under what must have been 20 coats of paint. I could just barely make it out, but it was there for sure. *In-Air-Rid*. I smiled.

There was a time in heating history when there was this huge company called American Radiator. They built a black skyscraper at 40 West 40th Street in midtown Manhattan (now a landmarked building). The building's roof is crenulated and painted with gold leaf, made to look like the glowing embers in a coal-fired boiler. It's lovely.

American Radiator made just about everything that had to do with heating, and they published these little red handbooks every year or so, which they called, *The Ideal Fitter*. I have a stack of them on my office shelf;, the oldest dates back to 1900. These are wonderful books to have because they show so much detail about the products that the American Radiator Company made back then. They describe the purpose and the inner workings of all those oddball gizmos that we find in steam-heated buildings. There are cutaway drawings of the products, and when you know what you're looking at, well; it just gets easier.

Doctor Bob's In-Air-Rid air vent didn't look like an air vent. It looked like a hexed plug, and it screwed into the last section of the radiator, right there at the top. From the outside, you'd never know this thing was an air vent. But look in *The Ideal Fitter* and you'll see how this wonderful device worked. All the inner workings of a normal air vent are there, but they're *inside* the radiator rather than outside the radiator. Behind the vent's float there's a spring-loaded, metal seat that pushes against the upper push nipple between the last- and the next-to-last radiator sections.

In a one-pipe-steam radiator, the steam enters from the bottom and displaces the air by rising above it (steam is lighter than air). The steam heads across the top of the radiator, moving through all the push nipple ports until it reaches the last section. If the air vent is at the high point of the radiator it will shut before most of the air has had a chance to escape from the radiator. That can lead to uneven heating, and it's the reason why the Dead Men installed their one-pipe radiator vents about halfway down those last radiator sections.

American Radiator got around this by fitting the In-Air-Rid with that spring-loaded seat. By sealing that internal opening, the seat forces the steam to work its way down that penultimate radiator section into the bottom of the last section. From there, it rises to the vent and all the air leaves the radiator. It is ingenious in its simplicity.

"Got a paperclip?" I asked. Doctor Bob looked around and found one.

"What are you going to do?" he asked.

"Watch this," I said, and then I straightened the paperclip and poked the thin wire directly at the dot on the letter "i" in the word "Air" in In-Air-Rid. The paperclip went right through the paint and into the radiator. "That's the vent hole," I said as we listened to the air escaping. "Beware of painters."

"Amazing!" Doctor Bob said. "All you did was poke."

"Yes," I said, "but it's not the poking; it's knowing *where* to poke that matters."

Dentists understand that.

A Lost Art revisited

My father retired from that job at the manufacturers' rep in 1985 and I left four years later to write a book about steam heating. I had written a small book for the company, which we called *The Steam Book*. It was a compilation of essays culled from a monthly newsletter that I had been writing for them for several years. We called the newsletter, *The Problem Solver*, and sent it to about 5,000 contractors and wholesalers each month. I used *The Problem Solver* to tell the stories about my experiences in the field. I wrote in the corporate "We," as in "We were on an interesting steam job with a contractor the other day, and here's what happened." I wove some heating history I had learned into the stories and our customers liked that a lot, so *The Steam Book* seemed like a natural extension. After all, this was New York, and there's plenty of steam heat in New York. The book didn't have my name on it, though, because my boss said people would bother me if they knew I had written it (I was young). They sold thousands of copies in the New York/Metro area, and that's what gave me the itch to become a *real* writer.

So I started a business with my wife, The Lovely Marianne, and set out to write a book I would call, *The Lost Art of Steam Heating*. I figured it would take me about six months to complete, but it took three years because the more I wrote, the more I learned. And this was before the Internet; I was getting all of my information from old books that I found in used-book stores around the country. The Dead Men would tell me their stories, and often refer to other Dead Men I had not yet met. So I would look for the books of those new Dead Men, and each filled in a piece of the puzzle that steam heating can be.

One day, it all came together for me, and I finished *Lost Art* in a flurry and we published it ourselves because we didn't think there would be a huge market for a book about 19th-Century heating technology. I was wrong. Our first printing of 5,000 copies of *The Lost Art of Steam Heating* sold out within six months, and it sold in every state (including Hawaii). Apparently, steam heating is everywhere. Who knew?

My biggest delight came on the day we received a purchase order for the book from NASA, which prompted The Lovely Marianne to say, "Hey, this isn't rocket science." Even NASA has buildings with steam heat.

Lost Art is now in its 20th printing and it continues to sell well because steam heat is so durable and so simple. It sticks around because it's difficult (and expensive) to rip it out and start anew.

How long will it all last? I suppose until the government outlaws it. They did that in Europe by limiting the maximum temperature at which a heating system may operate. Most of Europe heats with small, hot-water boilers that hang on the walls in kitchens, or with district hot-water heating systems, which take care of small towns. Europe had the "advantage" of starting over again after World War II. Our older buildings, those with the steam heat, were still standing after the war, and I can't imagine the politicians will get together anytime soon to outlaw those systems, but time will tell.

In the meantime, let's talk about what we can do to get the most out of these old systems. How we can make them use less fuel. How we can make them quieter. How we can get them to provide us with greater comfort. How we can we make them green.

And what the heck does *green* mean?

Green

In 2009, Lake Superior State University issued their 34th Annual *List of Words to Be Banished from the Queen's English for Mis-use, Over-use and General Uselessness*. At the top of the list was the word, Green, which made me smile since I'm using it as the theme of this book. Here are some of the comments from the folks who nominated "green" for the list:

"This phrase makes me go green every time I hear it." - Danielle Brunin, Lawrence, Kansas.

"I'm all for being environmentally responsible, but this 'green' needs to be nipped in the bud." - Valerie Gilson, Gales Ferry, Conn.

"Companies are less 'green' than ever, advertising the fact they are 'green.' Is anyone buying this nonsense?" - Mark Etchason, Denver, Colo.

"If something is good for the environment, just say so. As Kermit would say, 'It isn't easy being green.'" - Kevin Sherlock, Hiawatha, Iowa.

"If I see one more corporation declare itself 'green,' I'm going to start burning tires in my backyard." - Ed Hardiman, Bristow, Va.

"This spawned 'green solutions,' 'green technology,' and the horrible use of the word as a verb, as in, 'We really need to think about greening our office.'" - Mike McDermott, Philadelphia, Penn.

Ouch. And here we are "greening" steam. But in this book, the "green" that I'm going to be talking to you about is the green that goes in your wallet. Sure, fixing what ails any old steam system is going to be good for the environment because the boiler will burn less fuel and people will be less likely to open the windows by way of controlling the temperature in the rooms. But *saving* green (as in dollars) is what it's all about these days as the price of fuel creeps higher and higher.

So *that's* what we'll talk about.

Why now?

When it comes to saving dollars on heating, I see those old steam systems as delicious low-hanging fruit. There is so much you can do to make them better, and most of what you do won't cost a fortune.

During the summer of 2008, the price of heating fuel and gasoline skyrocketed and that got everyone's attention. We were pulling out the bicycles and looking into hybrid vehicles. But then the prices came down and many of us went back to our old ways.

But do you seriously think the price of fuel is going to stay low?

I don't either.

Later on, I'm going to tell you about a job that my friends, Frank "Steamhead" Wilsey and Gordon Schweizer did. They added some air vents and made a few other changes in a 32-unit apartment building, built in 1919. When they were done, they had cut the building's annual $35,000 heating bill by 32 percent.

Impressive, isn't it? That's low-hanging fruit. You don't have to rip out that old heating system and start anew. With steam, there is plenty you can do with a relatively small budget, and the results can make your jaw drop. Just ask Steamhead and Gordon.

If you're a heating contractor, this is a great time to be talking about greening steam because your customers are looking at their fuel bills and gasping. If you're a building owner, you should be talking to that contractor.

There's also federal stimulus money out there right now, and the low-hanging fruit is going to attract much of that money because a small investment can show a *very* large return on investment.

Just ask "Steamhead" and Gordon.

A story for you

Years ago, in some dank basement, a contractor installed a large air vent near the end of a steam main. He did this because he knew that steam and air are both gases, but steam is lighter than air, so the two won't mix. He also knew that the empty steam main was going to start out filled with air (which is only natural, right?), and that once it left the boiler, the steam would shove the air ahead of itself and down the main, just like a big, hot plunger. The contractor sized the air vent to release the air from the pipe at a certain rate so that the steam could move faster than he could run. The design velocity for steam in a heating system could be as high as 60 miles per hour. The only thing standing in the steam's way is air.

The contractor piped that vent and started the system, and it worked just as planned. But he wasn't surprised; he did this work every day, and he knew his business. He knew how to "think" like steam and like air. He knew that the air wanted a way out of the pipe, and that the steam was more than willing to give it a shove. That's why the air vent near the end of the main was so important, and why the contractor had to size it properly.

Time went by and the contractor eventually became a Dead Man. For many years, the steam plunged that air toward that vent, and the air left the system exactly as planned. The steam gave up its latent heat to the radiators and condensed into water, shrinking to $1/1,700^{th}$ the size it had occupied as steam. That's the expansion ratio between water and steam: 1,700:1. Think of it; one glass of water becomes 1,700 glasses of steam. And vice versa.

As the steam shrank back into water, a vacuum tried to form inside the steam main, but Mother Nature abhors a vacuum, and by this point, that big air vent had reopened, allowing the air back into the steam main. Every steam heating system ever built does this. It breathes out and it breathes in, just as we do.

When the air returned, the inside of the pipes was dripping wet from the steam. The oxygen in the air reacted with the wet metal, causing it to rust over time. Every steam heating system ever built rusts from the inside because every steam system ever built is open to the atmosphere.

The steam kept coming and coming and it moved so quickly that it carried with it the particles of rust and other solids that made its way into the system with the fresh feed water. Much of that crap wound up inside that big main air vent. It got stuck between the pin that closes the vent and the hole that releases the air from the vent. That's when the vent started to leak steam and water. It began slowly and no one noticed at first, but then it got worse and someone called in a new contractor.

This contractor hadn't grown up installing steam systems because people were now using hot-water systems in new buildings. The steam business had become one of replacing boilers and fixing parts. The new contractor had never seen an air vent like this one, so he removed it and took it down to his local supply house. He showed it to the counterman, who also had never seen one like this one.

"What's the problem?" the counterman asked.

"It's squirting steam and water all over the place," the contractor said.

"I know how to fix that!" the counterman said, and he handled the contractor a ¾" pipe plug.

"I've got these on the truck," the contractor said.

"So go use it," the counterman replied, and that's what the contractor did.

Now think like steam.

Back on the job, the steam was still trying to push the air from the vent, just as it had been doing for so many years. But now the vent was gone and there was a plug in its place. Plugs do a very poor job of venting (hence their name), so the air just collected down there near the end of the main, like traffic on a closed road. But the steam was flying out of the boiler. It didn't know the road was now closed. It was just trying to go to work as usual. It kept pushing and pushing, and pretty soon, the air had enough. It turned around and pushed back. Gases will do that. Try to compress them and the pressure will build. Scientists and engineers call this Boyles Law.

The people in the building weren't happy because a good part of the building was now cold. This is because where there is air, steam will not go.

So they called the contractor.

The contractor returned and felt the steam pipe. It was piping hot up to the point where the steam could reach, but a foot or so down the line, it was cool to the touch. This was because that part of the pipe contained nothing but air, and where there is air, steam will not go (are you getting this?).

"The pipe must be clogged," the contractor muttered to himself, and it *was* clogged, but with air, and few people think about air because they can't see it.

"Let me raise the steam pressure and see if I can blow this clog free," the contractor muttered. He went to the boiler room and adjusted the device that controls the pressure (we call this a "pressuretrol"), raising it from two-psi pressure to five-psi pressure. Then he waited 10 minutes or so.

To his delight, the steam had moved further down the line! This was because of Boyles Law, and also because of the First Law of Steam Heating, which states, *When you do something stupid, you will always get a reward, which leads you to do things of even greater stupidity.* By raising the steam pressure, the contractor had allowed the steam to further compress the trapped air. The air still had nowhere to go because that ¾" pipe plug was still doing what its name implies, but from the contractors' point of view, it looked like progress.

"Let me raise it a bit more," he mumbled.

Now he had the pressure up to eight-psi and the boiler ran almost all the time, burning lots of fuel as it did. The air was jammed up against that pipe plug and the steam reached further into the main, but never got to the end. He told the customer that this was the best they could expect. "It's steam," he said. "It's old."

The customer nodded and let it go at that. The following month, when the customer received his fuel bill, he gagged. It had never been this high.

You know what? You can move with a pinhole what you can't move with a ton of pressure. A simple main vent, properly sized, would have solved this problem and lowered that fuel bill. Steam is just like a delivery van. It moves down the road quickly, and it carries a full load of heat in its cargo bay. But if the road is closed, the van will just sit in traffic along with the rest of the cars and trucks and the heat will never get to the customers who are anxiously awaiting delivery. And they're paying for it.

If you "think" like steam and like air, you'll be able to see this, and you'll be able to hang onto more of that green because the key to saving fuel is to have the burner shut off. A burner that is off is 100% efficient. But before you can shut it off, you have to deliver the goods, and it's always best to do it in a way that the Dead Men intended.

For instance:

Steam heat should be silent

Water hammer, that banging and knocking that is so prevalent in steam heating, is not normal. Don't let anyone tell you that it is. A steam system shouldn't be the alarm clock for the building it serves. If the knocking and hammering is waking the tenants in the morning, or keeping them from getting to sleep at night, they'll eventually move out, and that's going to cost the building owner money.

Water hammer is also destructive. It can break pipes from fittings and cause injury. It's *not* normal. It has definite causes, and you *can* make it go away.

Hissing air vents aren't normal. In a well-balanced steam system, you should never hear air venting. When a vent makes noise, it's trying to tell you that it's handling too much air. High-velocity air makes noise. Want to hear it. Blow up a balloon, pinch the end and let the air escape. Noisy, right? That's the sound of high-velocity air. It's not normal for vents in a steam system to make that sort of noise. And when they are making that noise, debris from the system is racing into them, which shortens their life and costs money. Later on, I'm going to show you how to make noisy air vents be quiet, and that's going to save some green.

Screaming tenants *may* be normal, but they shouldn't be screaming about the noisy steam system. Screaming causes them to release large amounts of carbon dioxide into the atmosphere and that's not good for the planet.

And if they're screaming at you because you're a contractor, you're going to get stressed, and that's not good for you.

Steam heating, and tenants in steam-heated buildings, should both be silent. More on this as we move along.

Steam heat should be even

This is one of the biggest challenges for most people who own or work on steam systems, but you're going to see that it's not that complicated. It is possible to balance one- and two-pipe steam systems, and once you get them set up properly, some people will stop opening the windows to let out the heat. Others will stop yelling that it's too cold in their rooms. Balance leads to happy tenants and lower fuel bills.

Here are the main points we'll be looking at when we talk about making the steam heat even in every room:

First there's the water quality. If the water is fouled or not at the proper pH, the quality of the steam won't be good. And poor-quality steam won't travel very far. It quickly turns back into water before it has a chance to reach the places where the chilly people are.

Then there are the steam traps. You'll find those on most two-pipe steam systems. Traps set up the points of high pressure and low pressure, and steam will always move from high to low. When the traps fail (and they always will), steam gets into places where it doesn't belong, and that can really mess up distribution, making some rooms hot while others stay cold.

We'll also spend lots of time chatting about the right way to select air vents so that the steam goes down the mains and up the risers quickly and evenly.

So that's my introduction for you. Are you ready to dig into the nitty gritty of greening steam? Are you ready to learn how to save money on fuel?

Let's do it.

CHAPTER 2

GREENING THE LOAD

Meet John Henry Mills

One of the great-granddaddies of steam heating was John Mills, who worked with the H.B. Smith Company as a freelance inventor and engineer from 1873 until 1897. Between 1888 and 1890, he wrote a two-volume book titled, *Heat, Science and Philosophy of its Production and Application to the Warming and Ventilating of Buildings*. This magnum opus became an important resource for boiler- and steam engineers in the years that followed. He invented the Mills boiler, which Mestek, the successor company to H.B. Smith, still makes to this day. He came up with the idea that, in a tall building that was to have a one-pipe-steam system, it would be best to send the supply main straight up to the top of the building, and then downfeed all the radiators so that the steam and the condensate traveled in the same direction. He called this the Mills System and it worked beautifully. He also developed a quick method for figuring out the steam-heating load in a building, which folks called (not surprisingly) the Mills Rule.

Contractors loved the Mills Rule because it was so easy to use. They also began calling it the 2-20-200 Rule and I'll explain why in a minute. Most of the size-it-quickly rules of thumb that followed John Mills' rule evolved from his method, which were fine for their time but not so good nowadays because we build better buildings than they did.

John Henry Mills

Consider what sort of windows they used during John Mills' era. Most likely, they were single-pane, double-hung windows, with leaky sashes. There were no storm windows. And what about insulation? Do you think they used fiberglass batts back in the day? Do you think they used anything at all inside the walls? Not from what I've seen. The heat loss of a building was much greater back then, and the Mills (2-20-200) Rule was appropriate for that sort of construction. Not so good now.

Here's what I mean. Take any building and size a new steam boiler using the Mills Rule. The first thing you're going to do is measure the square footage of all the glass in the building. Once you have that number, divide it by 2 (that's the **2** in the **2**-20-200 nickname for the Mills Rule). Okay, now set that aside for a moment.

Next, measure all the cold surfaces in the building. A cold surface is any wall, floor, or ceiling that doesn't have heat on its other side. In a two-story house, the first floor walls are cold surfaces if they face outdoors. It's warm on one side of those walls and cold on the other. If a wall faces another heated room, you wouldn't measure that wall for heat loss.

You'd probably measure the ceiling on the second floor of this building because that ceiling faces the attic space (which is unheated), but you wouldn't measure the ceiling on the first floor of that building because that faces the heated second floor. The same goes for the floor; you'd count it if it was over a cold crawlspace or an unheated basement, but not if it was over a heated basement. Make sense? Good.

Okay, once you've measured all of the cold surfaces, divide that number by 20 (that's the second number in the 2-**20**-200 nickname). Put it on the back burner for a minute; we've got one more measurement to make, and this has to do with the air that's inside the building. The air is constantly changing because of infiltration. Old buildings were drafty (many of them still are). Measure the cubic feet of air by multiplying the length, times the height, times the width of each room. Now add those numbers and divide the total by 200 (the third number in the 2-20-**200** nickname of the Mills Rule).

The grand total you come up with will be the required square footage of Equivalent Direct Radiation (or E.D.R. for short) for the building. One square foot of E.D.R. for steam will give out 240 Btuh when there is 70-degree air on the outside of the radiator, and 215-degree steam on the inside of the radiator. That temperature of the steam is significant because 215-degree steam is steam at about 1-psi pressure, so what the definition of E.D.R. is telling us is you don't need pressure greater than 1-psi inside the radiator, even on the coldest day of the year. I'll have more to tell you about this later on when we get to the Greening the Pressure chapter.

Now here's the problem with the Mills Rule. We've upgraded the windows and even the insulation in many of those old buildings. The radiators are now oversized, based on the current heat loss of the renovated building. That can cause money to flow out through those new windows if people are going to be cracking them open so they can be comfortable. If you use the Mills Rule to figure out the radiation for a modern building, you'll probably wind up with enough radiation to heat *three* buildings. That's the problem with using sizing shortcuts from more than 100 years ago. They don't keep up with the times.

A footnote on Mr. Mills: Sometime in 1905 or 1906 (and this was in Westfield, Massachusetts), Mr. Mills wandered into town, dressed shabbily and looking penniless. J.R. Reed, who ran the H.B. Smith Company in those days saw him and said, "John Mills, I always warned you of this. Didn't I say that if you kept on at the rate you were going that you would surely scratch a poor man's pants?" He then gave Mr. Mills a check for $5,000 and said, "You are not going to give this money away or use it for any more experimenting."

John H. Mills never again appeared in Westfield. He died in 1908.

Here's to the Great Experimenter!

How much is enough?

The boiler's ability to make steam has to match the system's ability to condense steam. They *must* have equal strength. It's similar to an evaporator and a condenser in an air-conditioning system. The boiler evaporates the water by turning it into steam, and the Btus travel with the steam out into the system. The piping and radiation condense the steam by turning it back into water, wringing out the Btus as it does this. It's a simple process, but there must be a balance between the two.

If the "evaporator" (the boiler) is larger than the "condenser" (the piping and radiation), the burner will short-cycle because the system can't condense the steam that the boiler is making quickly enough. Short-cycling wastes fuel because the burner never gets a chance to settle into a steady-state condition, which is where the highest combustion efficiency resides.

On the other hand, if the "condenser" (the piping and radiation) can handle more steam than the "evaporator" can make, the burner will run on and on, always trying to satisfy the condenser's need to turn steam back into water. The condenser is voracious. No matter how much steam an undersized boiler makes, there never seems to be enough to reach the furthest radiators. Oh, it will eventually get there because of the boiler's built-in piping-Pick-up Factor (more on this later), but at what cost? An undersized steam boiler will use more fuel than a properly sized steam boiler. It's smaller, but it uses more fuel. Crazy but true.

So if you're a heating professional and you want to green steam, you'll have to match any replacement boiler that you install to the connected piping-and-radiation load. Don't even glance at the label on the existing boiler. Measure the load. That's what matters. Let's talk about the best way to do this.

Defining the load

The first radiator was a metal box that the famous James Watt (of steam-engine fame) placed in his home. The English loved their greenhouses, and James Watt had one. He warmed it with a steam kettle that sent steam through an open pipe directly into the greenhouse. The greenhouse itself became the radiator. The steam would warm the air directly and add lots of moisture, which the plants appreciated.

One day, James Watt got this idea. If he piped the steam into a metal box instead of just releasing it into the air, we could all have steam heat in our homes. Splendid! He just invented the radiator.

He let the air out of the metal box by opening a small valve at the top of the box (where there is air, steam will not go). It wasn't fancy or complicated, but it worked and every square foot of surface on that flat box was one square foot of radiation. Simple.

Time passed, and the rich people who could afford it took to this concept of central heating. Why be cold? But the rich folk wanted better-looking radiators, and manufacturers were happy to oblige. The radiators started to have pretty nooks and crannies, places where more air could touch more hot metal. This was when we began to see those lovely Victorian-style radiators with all their glorious metal embroidery. The radiators were part of the décor.

The term "square foot" of radiation was becoming more difficult to define at this point, and it was also becoming more difficult to measure. All those nooks and crannies! Sure, air had plenty of surface area to touch, but how were they to measure that surface area?

They solved the problem in a very ingenious way. They plugged all the holes in the radiator whose surface area they were going to measure, and then they submerged the radiator into a vat of paint. They first weighed the paint to see how much they had. They next removed the radiator from the paint and weighted the paint that was left in the vat. Whatever wasn't in the vat was now on the outer surfaces of the radiator. They took that much paint, by weight, and painted the floor. However many square feet of surface they were able to cover became the official square-foot rating of the radiator. This, they called Equivalent Direct Radiation, or E.D.R. for short. Pretty smart, don't you think?

E.D.R. became the standard term for measuring radiation and we still use that term to this day. For steam work, one square foot EDR will emit 240 Btuh, but only when there is 70° air on the outside and 215° steam on the inside. As I mentioned earlier, 215° steam is the temperature of steam at about 1-psi pressure. This is the key to understanding steam heating, and how to green it, and we'll look at it more closely in the next chapter.

The E.D.R. of any radiator will vary depending on the temperature of the stuff on the inside of that radiator, and if you were to convert a steam system to hot water, expect to feel less heat coming off the surface. The hot water isn't as hot as the steam.

The E.D.R. of any radiator will follow this chart:

Average temperature inside radiator (°F)	Output in Btuh per square foot EDR
215°	240
210°	230
205°	220
200°	210
195°	200
190°	190
185°	180
180°	170
175°	160
170°	150
165°	140
160°	130
155°	120
150°	110

Notice how when the temperature of the stuff inside the radiator drops by five degrees (Fahrenheit in this case), the output drops by 10 Btuh per square foot. It will keep doing that, all the way down the line until you can't feel any heat at all.

With steam at 215-degrees, you're getting 240 Btuh per square foot E.D.R. If you convert that same radiator to run on hot water, with an average water temperature of 170 degrees, the output will drop to 150 Btuh per square foot E.D.R. It's like turning down the heat on a stove.

So if you convert from steam to hot water, will your radiators be big enough? That's a good question to ask before making a decision.

Understanding boiler ratings

Boiler manufacturers use three ratings for steam- and hot-water boilers. The first rating, which they call **Input**, is easy to understand. It's the total Btuh value of the fuel that you're burning. It's what you're putting in. For instance:

No. 2 fuel oil yields 140,000 Btu per gallon

No. 4 fuel oil yields 155,000 Btu per gallon

No. 6 fuel oil yields 153,000 Btu per gallon

Natural gas yields 1,000 Btu per cubic foot, or 100,000 Btu per Therm

Propane yields 2,250 Btu per cubic foot, or 92,000 Btu per gallon

Electricity yields 3,415 Btu per Kilowatt hour

Anthracite coal yields 13,000 Btu per pound

Bituminous coal yields 12,500 Btu per pound

In the early days of steam heating, the ratings were more complicated because nearly everyone was burning coal, and coal wasn't as predictable back then as it appears to be on that list. Its heat value varied by the type of coal, the size of the pieces, the quality, the place where they mined it, and the amount of air the operator allowed into the boiler. It wasn't an automatic fuel, and that made a difference in the way the steam boiler operated. It often led to problems and misunderstandings.

During the 1950s, when contractors converted many of the older coal-burning steam boilers to burn fuel oil or natural gas, they included a very healthy safety factor – just in case (this was all new to them back then), and that's another reason why you're better off measuring the radiators when it comes time to replace that old steam boiler. To be green, the boiler's ability to produce steam must match the system's ability to condense steam.

Input is the heat that you're putting into the boiler. Not all of it is useful because no steam boiler operates at 100% efficiency. Some of that heat is going to go up the flue and out through the boiler's insulating jacket. What's left after those losses, and what's available to heat the building brings us to our next rating.

This is the **D.O.E. Heating Capacity**, or what they used to call the **Gross Output Rating**. D.O.E., in case you haven't figured it out, stands for Department of Energy. The government has their hand in these ratings because they also want you to be green. This rating tells us the amount of heat that we have available for the pipes and the radiators. It's what's left after the flue- and jacket losses.

Think about what a big job this leftover steam has to do. It not only has to heat all the radiators all the way across on the coldest days of the year; it also has to heat all the pipes between the boiler and the furthest radiator. That's tons of iron and steel, and it all starts out cold. The steam has to bring all that metal from ambient temperature to 215-degrees, and that's going to take some doing. If you do a good job of sizing, there will be enough steam left once all the pipes are hot enough to satisfy the radiators, which brings us to this next (and final) rating. This is the one that matters most.

The **Net Output Rating** is there for the radiators. That's what's going to heat the people in the building. And as you're thinking about this, consider what would happen if pipes that are covered with asbestos and all tucked in suddenly find themselves without that insulating jacket to keep the steam moving toward the radiators. What happens if the pipes *become* radiators?

We'll look more closely at that in a little while. It's real important, but first, let's talk about this thing we call the Piping Pick-up Factor. It's also real important, and often misunderstood.

Pick-up Factor

The Pick-up Factor represents the difference in load between the D.O.E. Heating Capacity Rating and the Net Output Rating. These days, the Pick-up Factor for steam is 1.33. That means that if you take the actual E.D.R rating of all the radiators you measure during your survey of the system, and multiply that by 1.33, you should get the boiler's D.O.E. Heating Capacity rating. So we're talking radiation, plus an additional third added on to the load for the pipes. That's what the Pick-up Factor is today. It wasn't always that way.

In 1940, for instance, the factor that boiler manufacturers used to designate the difference between a boiler's Net Output rating and its Gross rating (remember that's the old name for D.O.E. Heating Capacity) was 1.56. If you were working back then, you were probably the original installer. You would figure out the building's heat loss and select your radiators to overcome that heat loss. Next you'd take that installed radiation load in Btuh (or Square Feet EDR) and multiply it by a factor of 1.56 to get the boiler's Gross Output rating.

That's a pretty hefty increase, isn't it? You know why they did it that way? Because they weren't sure what they were doing. All of this was so new to them, and they were also dealing with gravity hot-water systems, which contained an enormous amount of water. They'd have to heat all that water on start-up, and that called for a lot of boiler. In those days, both steam- and hot-water boilers used the same Pick-up Factor of 1.56. By 1945, though, hot-water heating systems had changed. The pipes had gotten smaller and the systems didn't contain nearly as much water as those old gravity systems had held. So those in charge of such things decided that they were probably being too conservative. They reduced the Pick-up Factor for both steam- and hot-water boilers from 1.56 to 1.33.

Then, in 1967, they further reduced the Pick-up Factor from 1.33 to 1.15, but only for the hot-water boilers. They let the factor for steam boiler sizing remain at 1.33 because steam piping is larger than hot water piping; it has to be. Steam is bigger than water. Once again, to save the most green, the boiler's ability to produce steam has to match the system's ability to condense steam.

But what if someone has removed radiators from the building? It was too hot in there, so they stored some of the radiators in the basement. They didn't change the piping, though, and that's going to present us with a challenge. We now have more piping than we need for the radiators that are left, and that piping can condense steam, so we have to figure it into the Pick-up Factor.

Here's what I do when I see that radiators are missing (and I'll know they're missing because I'll see capped pipes). Instead of using the standard 1.33 Pick-up Factor that the boiler manufacturer has built into the D.O.E. Heating Capacity rating, I'll measure all the radiators, and multiply this by a new Pick-up Factor of 1.5. Then I'll select the boiler from the D.O.E. Heating Capacity column in the boiler manufacturer's catalog instead of from the Net column. This may give me a boiler that's a bit larger than it would be with the 1.33 Pick-up Factor, but it will be a boiler that can handle the extra piping that's still in place, even though the radiators are now stored in the basement.

It's an old-timer's trick and it will save you money by keeping that new boiler from running too long. To be green, the boiler's ability to produce steam has to match the system's ability to condense steam. Always.

How radiators heat

The steam goes into the radiator and the metal gets hot, right? The heat transfers to the air and there's peace in the valley.

Seems simple, I know, but there's more than that going on here, and *how* the steam enters a radiator affects the way people feel when they're near that radiator, so let's take a look at the different ways the Dead Men piped their steam radiators.

One-pipe steam

Steam is lighter than air. Just look at the way it leaves the surface of a pot of boiling water. It rises straight up, and it does the same thing when it enters a radiator.

One-pipe-steam radiators have, well, one pipe (hence the name). The steam has to enter at the bottom of the radiator because the condensate that forms after the steam gives up its latent heat to the cold metal and turns back into water uses gravity to drain back to the boiler. If you throttle the supply valve on a one-pipe radiator, the steam and condensate won't have enough room to get out of each other's way, so you'll get banging (which we call water hammer) and water will probably squirt from the air vent.

So one-pipe radiators are either on or off. That's it. They're either hot or they're cold. And that's a disadvantage. More on this later.

The steam enters from the bottom of the radiator and rises to the top. It moves from radiator section to radiator section, one at a time, pushing air ahead of itself as it goes. The air just wants to get the heck away from the steam. The air is looking for that vent, which will be in the last section of the radiator, the section that's furthest from the inlet valve, and part way down on the section (ideally, never at the top).

So one-pipe radiators heat from side to side, and it's normal for them to heat only part-way if it's not that cold outside. Once the thermostat (or whatever else is in charge of inside-air temperature) gets satisfied, the burner shuts off and the steam stops rising. One-pipe radiators generally get hot all the way across only on the coldest days of the year. That's normal.

Two-pipe steam

The best thing about a two-pipe radiator is that it has two pipes. Steam uses the supply pipe and condensate uses the return pipe, so

One-pipe steam radiator
Photograph by J. Crocker

they don't have to fight each other. There's less chance that the two-pipe radiator will bang. The air vent won't squirt (because two-pipe steam doesn't use radiator vents), and you can throttle the supply valve, which allows in a little steam or a lot of steam, all depending on how warm or cool you feel. This is *the* big advantage over the one-pipe radiator. It's no longer just off or on. You can set it so that it's just right.

And sure, there are more pipes with this system, and it was more expensive to install back in the day, but the pipes are generally smaller than the pipes in one-pipe-steam systems because each has to handle just one thing – either steam or condensate.

Steam can enter from the top or the bottom of a two-pipe radiator, which gave the installer some piping options. If the occupants didn't want to bend over to operate the valve, the installer could install the valve near the top. He could also have put it on the bottom and let them bend, or use a special foot-operated valve, which was popular during Victorian times when ladies wore those long dresses.

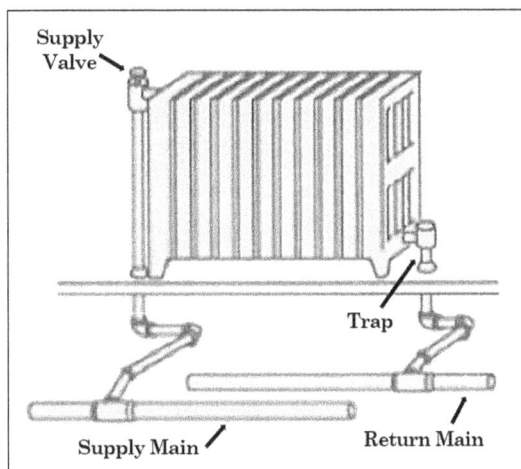

Two-pipe steam radiator

But whichever way the steam enters a two-pipe radiator, the condensate will always leave from the bottom because it's depending on gravity to get back to the boiler. It needs that low outlet to drain.

Now here's the best part. Steam is lighter than air, so if steam enters from the top of the radiator, it will try to stay there. It works its way across the radiator, condensing into water as it goes, and giving up both latent and sensible heat to the metal as the condensate dribbles down toward the radiator outlet. I'll tell you more about latent and sensible when we get to the chapter on Greening the Pressure, but for now, just think of all that hot stuff dribbling down inside the radiator. It's providing a nice, even warmth.

And if the steam enters a two-pipe radiator from the bottom, it will do the same thing. It will use the first section to rise to the top, and then it works its way across the upper push nipples (all two-pipe radiators have these), again heating the radiator from top to bottom.

I think two-pipe radiators give off a nicer, more even heat. It feels more radiant than what you feel coming off of a one-pipe radiator. I think that's why we see so many two-pipe radiators with top inlets. Keep the steam at the top and the radiator just *feels* better. Nice!

Two-pipe, air-vent radiators

This one's an oddball, but you may run into it, so I'll tell you about it. Before there were steam traps, which most two-pipe radiators have at their condensate outlet, the Dead Men used a supply valve on each side of the radiator, and an air vent on the side that has the return pipe. In this system, the radiator supply pipe is always one size larger than the return pipe. That gives the steam a path of least resistance to follow as it climbs the riser from floor to floor within a building.

If the building has just a few floors, the return lines probably drip directly into a "wet" (that means "below the boiler's waterline") return. This makes it impossible for steam to pass from one radiator to the next within that riser group, and it's the preferred way to do it. But in buildings that were tall, the Dead Men sometimes connected all the radiators to a common supply riser and a common return riser. The radiators became like rungs on a ladder and it was possible for steam to pass through the outlet of one radiator, say, on the first floor, and enter the radiator on the higher floors through *both* the supply and return lines. But the steam will

Two-pipe, air-vent steam radiator

always favor the supply line because it's larger than the return line and has less pressure drop.

The air leaves the radiator through the air vent, just as it will on a one-pipe radiator. In effect, this is a one-pipe system with drains.

If you want to put steam traps on this system, you'll have to add one to every radiator in the building. If you add a steam trap (or a thermostatic radiator valve) to just one or a couple of radiators, the steam will probably just enter the radiator from the return side, and either damage the steam trap or overheat the radiator that has a thermostatic radiator valve.

With these older systems, you have to think like steam. Where would you go if you were the steam? And is that the place you're supposed to be?

How to figure the size of a radiator

Okay, a boiler's ability to produce steam has to match a system's ability to condense steam. We're up to speed on that, so let's look at how we arrive at the size of the radiators. You'll need to do this whenever you're replacing an old steam boiler with a modern one. You're not sizing the boiler from scratch, as the Dead Man did, so you won't need to worry about the building's heat loss right now. All you have to do is provide enough steam to match what the radiation and piping can condense.

So get some paper and a pencil, a tape measure and a good flashlight, and then let's take a walk around the building. There are plenty of places to look for radiators, and we never know what we'll find.

Let's begin with these two basic types.

Column radiators

These were around in the 1800s and early 1900s, and they're exclusively for steam. Notice the big wide spaces inside each radiator section. That's to keep the steam's velocity and pressure drop low. Those wide sections make the steam feel welcome as it approaches. C'mon in!

Steam enters from the bottom of a column radiator and immediately rises to the top, just like oil on water. Notice that there are no nipples connecting the tops of these radiator sections. There's no need for a connection there because gravity does the trick. When I was first learning about all of this, I wasn't thinking about air as being heavy. I wasn't imagining it being able to fall (hey, air's light, right?), but then I started to think of steam as if it were helium, which is *lighter* than air, and I began to see it more clearly in my mind's eye. Steam rises; air falls.

It would be very difficult to use column radiators on a hot-water system because air would get trapped in the top of each radiator section. You'd have to drill and tap each section for an air vent, which would be difficult to do, and would look silly. Or, I suppose, you could try turning the radiator upside-down, filling it with water, and then flipping it over quickly. Be careful, though, because that radiator probably weighs more than a sumo wrestler at that point. Try not to spill a drop.

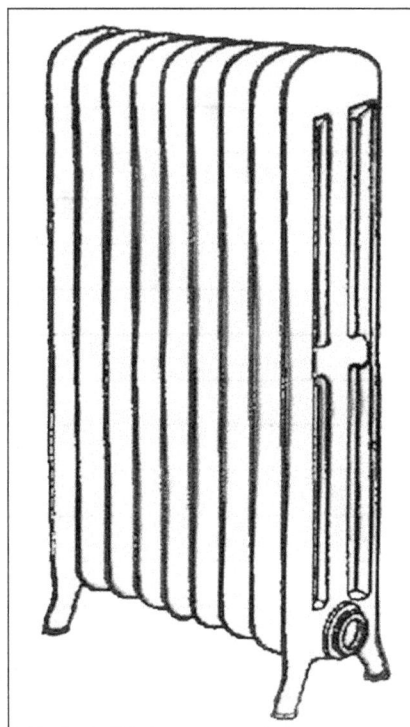

Column radiator

Just kidding.

Anyway, here are some basic E.D.R. ratings for column radiators. If you want *exact* ratings for a particular make and model, check out my book, ***E.D.R. - Ratings for Every Darn Radiator (and convector) You'll Probably Ever See***. There are some more ratings there.

A "column" is what you see when you view the radiator from its narrow end. For instance, in the picture on the previous page, the radiator has three columns and nine sections.

Number of Columns	Column Height (in inches)	Sq. Ft. E.D.R. Per Section
1	20	1.5
1	23	1.66
1	26	2
1	32	2.5
1	38	3
2	20	2
2	23	2.33
2	26	2.66
2	32	3.33
2	38	4
2	45	5
3	18	2.25
3	22	3
3	26	3.75
3	32	4.5
3	38	5
3	45	6
4	18	3
4	22	4
4	26	5
4	32	6.5
4	38	8
4	45	10

Oh, and when you're measuring the height of the radiator, go from the bottom of the radiator's foot to the top.

Thin-tube radiators

The Dead Men started using these radiators for steam at the turn of the 20th Century because these radiators have push nipples across both the bottom and the top of each section. This makes two-pipe steam possible. Vapor heating was becoming popular around this time ("vapor" means the high pressure is always less than eight ounces), and these radiators were just right for that. You get a nice, even radiant glow from these, and you can use these radiators for both steam and hot water.

Once again, measure the height of the radiator from the bottom of its foot to its top. And for purposes of this chart, a "tube" is what you see when you view the radiator from its narrow end. The radiator I'm showing you in the drawing has five tubes and eight sections. Go ahead; count them yourself to make sure I'm right.

Here's the basic chart:

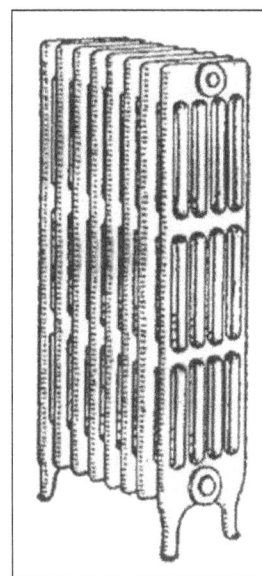

Thin-tube radiator

Number of Tubes	Tube Height (in inches)	Sq. Ft. E.D.R. Per Section
3	20	1.75
3	23	2
3	26	2.11
3	30	3
3	36	3.5
4	20	2.25
4	23	2.5
4	26	2.75
4	32	3.5
4	37	4.125
5	20	2.66
5	23	3
5	26	3.5
5	32	4.33
5	37	5
6	2.	3
6	23	3.5
6	26	4
6	32	5

continued from previous page...

Number of Tubes	Tube Height (in inches)	Sq. Ft. E.D.R. Per Section
6	37	6
7	13	2.625
7	16-1/2	3.5
7	20	4.25

Once you've found and figured out the E.D.R. for all the radiation in the building, you'll add a suitable Pick-up Factor (based on whether or not all the original radiators are still installed), and then use that number to select a replacement boiler, based on the boiler manufacturer's D.O.E. Heating Capacity. That boiler's ability to produce steam will match that system's ability to condense steam.

Reflective insulation

You'll sometimes see a thin, reflective insulation barrier between a radiator and the outside wall of a building. The Dead Men installed these to reflect the radiators' radiant heat into the room. Typically, a freestanding cast-iron radiator gives off about 40% of its heat by radiation and 60% of its heat by convection. These reflective barriers aren't thick enough to take the place of true inner-wall insulation, but they *will* direct the radiant energy that's flowing from the hot metal toward where the people are, rather than toward the cold wall. And the older those walls are, the less chance there is that they will contain any insulation.

In New York City, the Empire State Building is getting a green retrofit as I write to you, and one of the things they're doing to save energy is to add reflective insulation barriers behind the radiators (there are about 6,500 of them) on the building's outside walls. I think these barriers make sense. They're not expensive, they're usually very easy to install, and they're as green as can be.

Insulate the steam pipes - always

Asbestos (the Magic Mineral!) was a wonderful insulator. No insulation is perfect; heat will always flow toward cold, but asbestos did a fine job of tucking in the steam like a baby in a blanket. It gave the steam a chance to move from the boiler to the radiators without changing back into liquid water along the way. I can recall walking into boiler rooms where guys were tossing the stuff around like snowballs. "Hey, Dan!" they'd say, and then hit me right in the puss.

Who knew that stuff could kill you?

Nowadays, much of the asbestos is gone from the pipes, and many people with steam heat didn't bother to replace it with fiberglass insulation because they liked the way their basement felt without the insulation. It was cold when the asbestos was on the pipes, but now it's cozy down there. Here's why:

Let's say there's a 2-1/2" steam main running from this side of the basement to way over there. The pipe is bare because someone removed the asbestos. The heat loss per linear foot of that pipe is going to be 260 Btuh when steam is at 1-psi pressure inside the pipe. Take a moment to compare that to the Square Foot Equivalent Direct Radiation of a radiator. One Square Foot E.D.R. gives up 240 Btuh when there is 70° air on the outside and 215° steam on the inside. That uninsulated 2-1/2" pipe does an even better job, and it becomes a *very* effective radiator.

Okay, now let's put just one-inch of insulation over that pipe. The heat loss from the pipe immediately drops to just 49 Btuh per linear foot because now the heat is tucked in. With just one-inch of insulation, the pipe's heat output is less than one-fifth of what it is when the pipe is bare. This is why the basement with the bare pipes is so cozy, and the basement with the insulated pipes isn't as warm. The steam stays inside the insulated pipe and travels to the radiators, which are where the people are. And isn't that where you want the heat to be?

You can add more insulation to the pipes if you'd like, but the first inch is the most important. After that, the return on your insulation investment starts to diminish. There's always going to be some heat loss, no matter how much you insulate those pipes, but a thickness of one-inch is usually all you need for steam heating.

Straight pipes are easy to insulate and the insulation doesn't have to be fancy. You can buy the clamshell-type of insulation that fits snuggly around the pipes and seals with tape. Or you can just wrap the pipes with batt insulation and hold it in place with cable ties.

Some people don't insulate the fittings and valves, thinking that this isn't important, but it is. Fittings and valves have more surface area than pipes because of the nooks and crannies, and the heat loss from these can be nearly three times greater than it will be from a straight run of pipe of the same size.

Insulating fittings used to be expensive because you had to buy special form-fitted pieces, but now there are insulating blankets that you can fit around a fitting or a valve and tie in place. This makes it easy to get at those fittings and valves later on. You can remove the blanket and reinstall it afterwards without damaging it.

Like any insulation, pipe insulation works by trapping layers of air, making it difficult for the heat to move from the steam to the surrounding area. Air is a lousy conductor, but water isn't, so make sure that the insulation doesn't get wet, and if it does get wet, replace it with dry insulation, which will pay for itself quickly in saved Btus.

Oh, and if you like to climb around commercial boiler rooms, please don't step on the pipe insulation. If you crush it, it's going to be a lot less effective because it won't contain nearly as much air.

Finally, the nicest thing about pipe insulation is that you usually only have to buy it once. Fuel, on the other hand, you buy over and over again. Without insulation, much of the steam will decide to stay in the basement, making it cozy. It's not going to be upstairs where the people are. They'll keep touching the thermostat, which will keep the burner running. And that's costing you some serious green.

Why steam radiators are so large

For starters, check out the insulation in those old buildings. There probably isn't any. Or maybe there's some straw and mud within the walls. And how about those windows? Not so good, are they?

Lots of heat loss means lots of radiators (and big ones!), but there was more than that going on during the early days of central heating, and this, too, had an effect on the size of the radiators we have to deal with today. Here's the story:

During the winter of 1866-67, just after the Civil War, Lewis W. Leeds gave a series of lectures at the Franklin Institute in Philadelphia, Pennsylvania. Mr. Leeds was an important guy at the time. He was the Special Agent for the Quarter-Master General for the Ventilation of Union Hospitals during the Civil War (some title, eh?). He was also the consulting engineer for all the heating and ventilation in the U.S. Treasury buildings. And this was when central heating with steam was brand-spanking new.

His *Lectures on Ventilation*, at the Franklin Institute had an ominous subtitle: "Man's own breath is his greatest enemy." Now, keep in mind that Mr. Leeds was around during the time before we fully understood that there are germs, and he believed that vitiated air caused most of the illnesses that killed people, especially young people. Harriet Beecher Stowe, of *Uncle Tom's Cabin* fame, got behind his theories, and that *really* helped spread the word.

People were just beginning to try central steam heating, and the thought of what might happen if all the windows were closed in a hot, stuffy room, scared them. So the engineers of the time oversized the radiators. Those radiators would have to be able to get the job done with the windows open.

Years later, The Spanish Influenza pandemic arrived, and when it took the lives of 675,000 Americans (and 50 million, worldwide) during the winter of 1918-19, it got everyone's attention. The flu spread through the air, and the last place anyone wanted to be was in a closed room with other people who were coughing and sneezing.

In 1919, the United States Board of Health directed people to keep their windows cracked open during the winter to avoid disease. That's another reason why most steam radiators are so big.

Oversizing radiators was normal during the Roaring '20s, but when the Great Depression arrived (and because the Spanish Influenza never returned), people began shutting their windows to save fuel. They roasted, of course, and this is where the government steps in with a neat solution. And while this solution didn't actually save any fuel (you'll see why in a moment), it did make steam-heated rooms more comfortable, and it even helped to balance those old steam systems.

On July 19, 1935, the U.S. Department of Commerce's National Bureau of Standards published a report explaining how paint that contains metal flakes could cut a radiator's radiant output by as much as 20 percent. It caught the public's attention and this is why most old radiators are painted either silver or gold. It's called "bronzing," and it's only the final coat that makes a difference in the radiator's output. It's not so much the color as it is the composition of the paint itself that makes it work, but I'll let you read it for yourself.

Here 'tis:

Painting of Steam and Hot Water Radiators

For a number of years this subject has received considerable attention from the public, and it is apparent that the essential facts have not always been understood. The object of this note is to supply the more important facts in the case.

It will appear that as far as their effect on the performance of radiators is concerned, paints fall into two classes. First, those in which the pigment consists of small flakes of metal, such as the aluminum and bronze paints, most commonly used for painting radiators, which produce a metallic appearance and will be called metallic paints. Second, the white and colored paints, in which the pigment consists not of the metals but of oxides or other compounds of the metals. Thus, white lead paints, or those containing compounds of zinc or other metals, will be called non-metallic paints. These non-metallic paints are obtainable in practically all colors, including white and black, while the metallic paints have the color of the metal or alloy of which the flakes are composed.

We will state at the outset the principal conclusion, which will be explained in more detail later, that the last coat of paint on a radiator is the only one that has an appreciable effect. And that a radiator coated with metallic paint will emit less heat, under otherwise identical conditions, than a similar radiator coated with non-metallic paint. In order to obtain the same amount of heat from the two radiators just considered the temperature of the one painted with metallic paint must be somewhat higher. Under these conditions, exactly the same amount of heat is being supplied to the two radiators. And since neither the boiler efficiency nor the heat wasted in the pipe lines is appreciably affected by small changes in radiator temperatures, practically the same amount of fuel is required to supply the heat in each case. In other words, while it may be desirable for various reasons to avoid the use of metallic paints on radiators, no appreciable saving in fuel will result from the use of non-metallic rather than metallic paints.

The purpose of a heating system is to maintain the rooms in a house at some temperature higher than that prevailing out of doors. The heat that is developed by burning fuel is transferred to the rooms by means of the radiators. A radiator neither creates nor destroys heat and a large radiator, while it can put more heat into a room than a small one, must be supplied with all of the heat it puts in. In the sense that they ultimately transfer all the heat supplied into the room, all radiators are 100% efficient. The word "efficiency" is, however, used in other ways, and it is now customary to use it on all possible occasions, but it is hardly correct to say that putting metallic paint on a radiator reduces its efficiency when the effect is merely to reduce its capacity. The size of the radiators in a house can only affect the fuel required for heating by increasing or decreasing the heat wasted in transmission from boiler to radiator and that lost up the chimney. Only when the radiators are so small as to render the whole heating plant ineffective is an appreciable saving of fuel to be expected by installing larger radiators.

After these preliminary explanations, we may proceed to consider the kind of effects that may be obtained by the use of various kinds of paint. The heat emitted from a radiator is removed in two ways. First, the air streaming past the radiator and rising from it is heated and carries the heat to other parts of the room. Second, the hot surface of the radiator emits heat by radiation just as the glowing electric and gas heaters do. Most types of steam and hot water radiators emit less than half their heat by radiation and evidently the name "radiator" although universally used is not a particularly appropriate one.

To take concrete case, a particular sectional cast iron radiator, if painted with any non-metallic paint, might transfer into the room 180 Btu per hour for each square foot of its surface, if supplied with the necessary amount of heat from a boiler. The burning of one pound of good coal produces about 12,000 Btu, and if the coal is used in a domestic heating plant, perhaps half of this, or 6,000 Btu, might finally be transferred from the radiators into, the rooms. Most of the other half of the heat produced is inevitably lost via the chimney.

The area of one section of a cast iron radiator is about two square feet for the smaller sections, and up to seven or eight square feet for the larger sections, so that a 10-section radiator would have a surface area between 20 and 80 square feet.

Of the 180 Btu per hour transferred, about two-thirds, 120 Btus, would go to heating the air that passes over the radiator. The 120 Btu transferred directly to the air would not be increased or decreased by repainting the radiator. The remaining 60 Btu not carried off by the air is emitted as radiant energy. The amount of radiant energy that can be emitted per hour by the hot surface is dependent upon the kind of paint used for the last coat. It was assumed that the radiator was painted with non-metallic paint. If it be repainted with a metallic paint, such as aluminum or bronze, it will no longer be able to radiate 60 Btu per hour, but may be able to radiate only 30 Btu, so that instead of transferring 180 Btu to the room per hour, it can now transfer only 150 Btu. The coat of aluminum or bronze paint is not an insulating covering like a covering of magnesia or asbestos, but it has a similar effect, although for an entirely different reason. The resulting reduction in heat emission is entirely due to the reduction in the radiating power of the exposed surface, rather than to the insignificant insulating value of the thin layer of paint. It is therefore evident that undercoats of paint, regardless of kind, have no significant effect on the performance of the radiator, except in the practically impossible case where the paint was thick enough to act as an insulating covering. In repainting a radiator, it is therefore unnecessary to remove the old paint. The effect of adding the metallic paint is equivalent to removing 1/6 of the radiator, or nearly 17%, or as if one section out of six had been removed. Thus, a radiator of five sections painted with white or colored paint should be about as effective as another of six sections of the same kind painted with metallic paint since each would transfer the same amount of heat to the room to provided the necessary amount of heat were supplied to each.

In the following applications, the numerical values given above will be used as if they were exact, but it must be understood that they are merely representative and would not apply exactly to any particular case except by chance. The effect of painting on the capacity of a radiator depends upon the size and design of the radiator. The reduction in capacity produced by the application of aluminum paint is less for large radiators than for small ones, especially so in the case of large radiators having many columns or tubes per section. In a large tubular type radiator having seven tubes per section, more than three-quarters of the heat is carried away by the air directly and painting with aluminum consequently reduces the capacity of the radiator only about 10%. If only the visible portions of a radiator are painted with aluminum paint, the reduction in capacity is also obviously less than if the entire surface is covered.

Application 1: *Suppose a house in which all the radiators are painted with aluminum paint and that the radiator in one room is found to be too small, so that when the other rooms are warm enough, this one is too cold. If the radiator in this room is painted with non-metallic paint, either white or colored, the heat emitted by it can be increased from 10 to 20% without affecting conditions in the other rooms, although it will be necessary to burn more fuel to supply the additional heat in the one room. If the increase is sufficient, the expense of installing a radiator may thus be avoided.*

Similarly, it is possible, by using bronze or aluminum paint on radiators in rooms which are overheated, and colored or white paints in rooms not sufficiently heated, to improve conditions without going to the expense of installing new radiators of larger or smaller sizes.

Application 2: *In installing radiators in a new house, somewhat smaller radiators may be installed if they are to be painted with colored paints, rather than bronze or aluminum paints.*

Application 3: *If the radiators on a hot water system are painted with metallic paint and are all too small, so that the water must be kept hotter than it is desired in order* *to heat the house, they may be repainted with non-metallic paint, and it should then be possible to heat the house with the water in the system not quite so hot. There will be no noticeable saving of fuel.*

Application 4: *Since basements are usually overheated so that much of the heat supplied there is wasted, some economy can be effected by painting the heater and pipes, with metallic paint. This cannot, however, serve as anything more than a poor substitute for a covering of good insulating material about one-inch thick; which is capable of making an appreciable saving in the coal bill. The insulating material will remain effective for years, while the paint becomes ineffective if covered with dust.*

Application 5: *If a radiator is situated next to an outside wall, as most of them are, it is evident that the heat supplied directly to this wall is more or less wasted. Some slight economy may be obtained, therefore, by using metallic paint on the side facing the wall and non-metallic paint on the visible portions. The gain is not large enough to be important, but on the other hand, in putting non-metallic paint over metallic, it is not worth while to go to the trouble of repainting' the side next the wall.*

Pick a color!

Here's a chart that shows the effect different colors and types of paint will have on a radiator's ability to radiate heat. Keep in mind; this doesn't affect the radiator's ability to pass heat to the air by convection. That's coming up next.

Surface color	Percent Effectiveness	Btuh per Square Foot E.D.R.
Cast Iron (not painted)	100	240
Terra Cotta	103.8	249
White Zinc or Enamel	101	242
Maroon Japan or Flat Black	100	240
White Lead	99	238
Green Enamel, Dull	96	230
Gold Bronze	81	194
Aluminum Bronze	80	192

Can you see why so many old radiators are painted silver? Check out that last rating. The silver paint cuts the radiators ability to radiate by 20%. It's a nice and easy way to cut the output of an oversized radiator.

Should you box those radiators?

Let's take a look at radiator enclosures. Some people use them because they think old radiators are ugly (I sure don't). Others use them to protect children from burns (I recall touching a red-hot steam radiator when I was a kid growing up in New York City, but I only touched it once). Still others think that by using an enclosure, they'll get more heat from the radiator, and this will be true if it's the right enclosure. The folks who market enclosures often claim that their products will increase a radiator's output, but that's not always true. Often, the enclosure will decrease the output. It all depends on how the air flows through the enclosure.

Here's a chart from the old days that shows what's going on.

Let's look at each drawing, starting from the top, left to right. First, we have a radiator with a solid board in front of it. That board is going to create a chimney effect for this radiator. The air that comes in contact with the hot metal will quickly rise, drawing in cool air from the bottom of the board. Because more air will come in contact with more hot metal, the chart tells us to deduct 10%. What that means is that if you have a room that needs a radiator capable of putting out, say, 100 Square Feet E.D.R., you could use a radiator rated at 90 Square Feet E.D.R. in this case because the board is increasing the air flow across the radiator. It's similar to what happens when you start a fan and allow it to blow across a radiator. More air flow means more heat output.

Okay, move to the right. This next radiator has a simple shelf across its top. That shelf is impeding the air flow off the hot metal, so we're going to have less convection with this one. If we needed 100 Square Feet E.D.R., we'd have to size the radiator for 120 Square Feet E.D.R. to compensate for the lesser convection. Make sense? (This could actually help you if the radiator is too big for the room.)

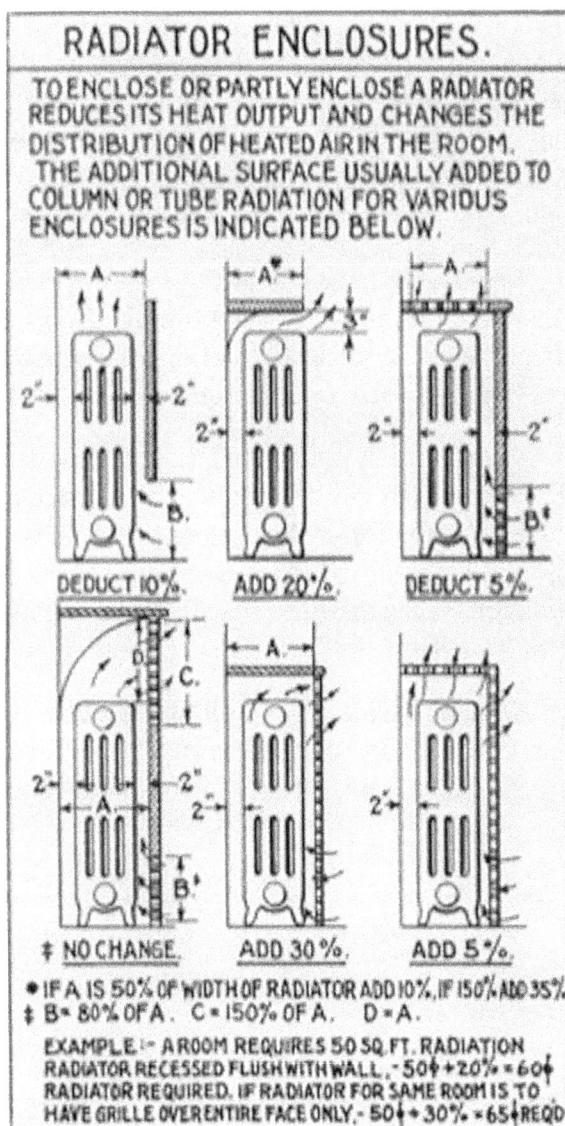

RADIATOR ENCLOSURES.

TO ENCLOSE OR PARTLY ENCLOSE A RADIATOR REDUCES ITS HEAT OUTPUT AND CHANGES THE DISTRIBUTION OF HEATED AIR IN THE ROOM. THE ADDITIONAL SURFACE USUALLY ADDED TO COLUMN OR TUBE RADIATION FOR VARIOUS ENCLOSURES IS INDICATED BELOW.

DEDUCT 10%.   ADD 20%.   DEDUCT 5%.

‡ NO CHANGE.   ADD 30%.   ADD 5%.

* IF A IS 50% OF WIDTH OF RADIATOR ADD 10%, IF 150% ADD 35%.
‡ B = 80% OF A.   C = 150% OF A.   D = A.

EXAMPLE :- A ROOM REQUIRES 50 SQ. FT. RADIATION RADIATOR RECESSED FLUSH WITH WALL, - 50⊕ +20% = 60⊕ RADIATOR REQUIRED. IF RADIATOR FOR SAME ROOM IS TO HAVE GRILLE OVER ENTIRE FACE ONLY, - 50⊕ + 30% = 65⊕ REQD.

The radiator to the right of that one has an enclosure that's similar to the first one. We have a solid front and a top and bottom that are perforated with lots of holes, giving air a way in and out. Because of this design, we can deduct 5%. So if you needed 100 Square Feet E.D.R., you'd be able to get by with 95 Square Feet E.D.R.

Oh, and this is a good point to mention that when you're measuring radiators for a boiler replacement, please ignore those enclosures in your measurements. You have no way of knowing whether someone will remove those enclosures as time goes by. Better to be safe than have a boiler that's suddenly too small or too large.

Let's move to the bottom row, starting on the left. This next enclosure is well-made and it neither slows the air nor speeds it up, so there will be no change in the radiator's output.

Moving to the right, we see the classic radiator enclosure. This is the one that you'll see lots of companies selling nowadays. It has a solid top that's usually hinged (so that when you open it you can see all the cobwebs and other crap that's fallen in there). It also has a perforated wood or metal front, drilled with hundreds of holes. There's little or no chimney effect with this enclosure so we have to add 30% to its size if we want the same output that we'd get without this enclosure. So if we need 100 Square Feet E.D.R., we'd have to size for 130 Square Feet E.D.R. And once again, *don't* add that to your measurements when you're sizing a replacement boiler.

Finally, we have this last one on the lower right. This is very similar to the previous one (there are lots and lots of holes in the front) but we've added holes to the top so that's going to let the hot air escape. You'll have to add 5% to this one; so to get 100 Square Feet E.D.R., you'll need to size for 105 Square Feet E.D.R.

And let me say this again: You are *not* the one sizing the radiation to the room's heat loss. The Dead Man did that years ago. He may or may not have sized for these enclosures. You have no way of knowing unless you do a heat-loss calculation on the room as it was in the old days (and how are you going to do that?). And then you'll have to check the size of the radiator against that heat loss. Trust me on this. If you're measuring radiators for the purpose of replacing the boiler, take the measurements as if the enclosures were not there. You'll be fine.

Oh, and notice how in all of these drawings, the radiator is positioned two inches away from the wall, and two inches away from the front of the enclosure. That's an ideal spacing for air flow. Keep this in mind if you ever have to move a radiator.

How do you paint steam radiators?

That's the question I asked the people who participate on The Wall (our very active forum at HeaingHelp.com). Here's what they had to say about their experiences:

John: When customers ask, I tell them to clean the radiator with trisodium phosphate, and then prime it with an alkyd, oil-based primer, followed by a good latex top coat. Sherman Williams and California are two top brands that I have used with good results. Use a brush and hot dog roller. Select a top coat that's a shade lighter than the color of the wall. Radiators seem to darken after a few heat cycles and then blend with the wall.

Mike: I am a painting contractor and we always use Benjamin Moore Satin Impervo to paint radiators. It is alkyd enamel that sprays well and has a beautiful finish. If we cannot convince the homeowner to remove the radiators so that we can properly paint them in our shop, we are forced to paint the radiators in place. We use either a hot dog roller or a paint brush designed to paint radiators (it's shaped like a hockey stick and you can find these with a Google search). When painting radiators in place, the oil-based enamel has excellent adhesion to the marginally prepared surfaces. Latex paint is for homeowners.

(So Mike, the professional painter, disagrees with John, the professional heating contractor on whether the top coat should be latex or oil-based. It sounds like they both get good results, though – D.H.)

Thad: Sandblast and powder coat for a couple hundred bucks per radiator and they'll look brand new. The real trick is to bake them for a bit longer so that any outgassing from the cast iron occurs before the powder coat dries. This will avoid any surface blemishes. I have had seven, 100-year-old steam radiators done this way and they all came out fantastic. And a little bonus is that the finish is so smooth that dust and cat hair don't get stuck in between the sections. It just comes out with a whiskbroom, which I use once a month.

Phil: I sandblasted one myself last fall and painted it with Rustoleum (their bronze metallic finish). The paint went on easily, dried quickly and the radiator looked terrific after painting. It did take a couple of days to finish outgassing once the heat came on, though, so some people will have a problem with that. The paint is somewhat soft, and will get tacky when it is hot (the kids hats and gloves tend to stick when left to dry) and it seems to collect dust and hair. I may try the powder coating method next time, and compare the costs (the do-it-myself solution cost less than $20).

(You get what you pay for. – D.H.)

Mike T.: I have only painted a few steam radiators, and for most of them, I used plain latex wall paint over an already-sound, painted surface. I saw one batch a few years later, and they were okay, but not great, with a few small areas of rust coming through.

I did a couple with traditional, oil-based silver "radiator" paint. The condition of the existing paint (silver) was not too good, but the customer didn't want to go the expense of stripping the old paint. I haven't seen it, but the customer never complained. I haven't yet found anyone willing to pay what I consider a very reasonable fee for one of my custom bronzing jobs.

Tim: *I sandblast them and then use an automotive spray gun for the paint. I did use Rustoleum on my first three last year. I had to thin it by half for it to work with the spray gun. I finally bought a compressor to handle the air requirements for the sandblasting before the painting. What a difference!*

Kevin: *I painted two radiators three years ago with Rustoleum high-heat paint in a spray can. They're still perfect. The can says the paint can handle up to 1,200 degrees, but there are not a lot of choices when it comes to colors.*

Dave: *I have radiators sandblasted and powder-coated all the time. They look stunning when they are done. I use a commercial painting company that handles both the blasting and the coating. They have many colors from which to choose. The radiators are cooked at 400-degrees Fahrenheit. It's important not to get them any hotter than that because the paper gaskets between the radiator sections can be damaged, causing the radiator to leak. Ask me how I know. A large radiator can easily cost as much as $400 for the whole process, plus the time it takes to disconnect, transport to and from the paint shop, and reinstall afterwards. The finish is quite durable, though, and it looks like porcelain.*

Patrick: *Ditto to everything that's been said about powder-coating. I hate the thought of paying someone else to do what I can do myself with a little sweat and elbow grease, but powder coating is so superior compared to the results I can achieve with even a professional spray setup that I find it's really worth the cost. I've had zero off-gassing issues; the finish is perfect (no drips, drabs, or missed spots), and let's face it – doing a good job of cleaning and painting a radiator of any size is one of the more onerous tasks imaginable. Refinishing radiators seems to me to be a textbook-perfect example of when powder coating makes sense.*

Sure sounds to me like powder coating wins. Check with a local auto-body shop. They'll often do this sort of work for you. And I've had pros tell me that they've taken radiators to monument makers at local cemeteries for sandblasting.

Painting radiators for aesthetic reasons may not make our planet any greener, and it *will* take some green out of your wallet, but when that old Victorian beauty is gleaming like new, it sure does make you smile.

And happiness also matters in this new green world of ours.

Indirect radiators

These are interesting radiators because they're large and they hide inside basement ductwork. The Dead Men chose this system because it combined heating with ventilation. The name "indirect" comes from the idea that the radiators aren't in the same room as the people. These radiators heat the incoming air, which then rises through the ductwork to the first, second, and sometimes even the third floor of a big house. They put together most of the indirect radiators as you would a small cast-iron, sectional boiler. Some, however, were made of rows of steel fin-tube. In each case, though, it's tough to figure out what size the indirect radiators are because you can't see them, but here are some tips that should help.

Any indirect radiator has to be at least 14 inches higher than the steam boiler's waterline. This is to allow for the gravity return of the condensate back to the boiler. Most of these systems worked on very-low-pressure steam, usually just a few ounces. If you're replacing an old boiler you have to be very careful where you position the new boiler's waterline because indirect radiators often hang very low. If you set your replacement boiler too high, it may partially fill the indirect radiator with water, and that will seriously cut down on the radiator's output.

Look around the indirect steam radiator and make sure the air can get out. Ask yourself that key steam-heating question: If I were air, could I get out? Look for an air vent on the outlet side of the indirect. The air has to make it completely through the unit if the steam is to arrive on time. I mention this because many of those vents are gone now. They leaked and some knucklehead replaced them with plugs. Plugs don't vent well.

Within the duct, the indirect radiator has to be about 10 inches below the top, and eight inches above the bottom. The radiators have to be tight against both sides of the duct. These dimensions are crucial to the proper flow of air across any indirect radiator. Sometimes, a cast-iron unit will fail and you might want to replace it with a homemade nest of fin-tube

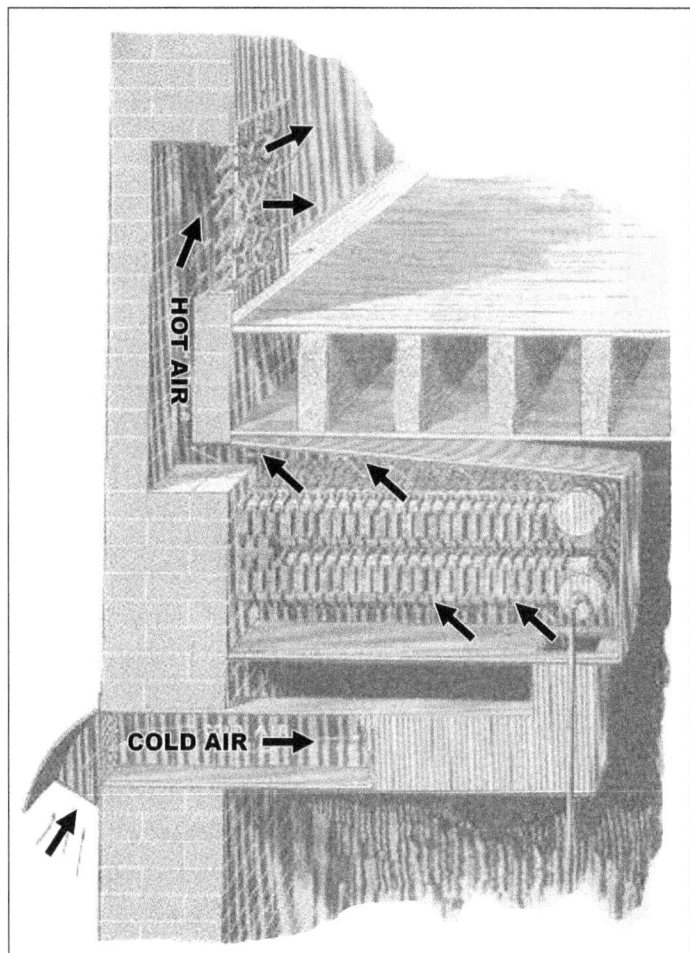

Indirect radiator

radiation. Watch what you're doing, though, because the flow of air is so subtle here, and so important to the unit's Btuh output.

When the Dead Men used the indirect radiators for ventilation, as well as for heating (which was most of the time), they always tried to get the outside air to enter from the bottom of the indirect radiator. If this wasn't possible, they took the next-best option, which was to bring the fresh air in from the side opposite the warm-air outlet. There are no return-air ducts in this system. The ventilation air enters the house without benefit of a fan. The only way it can do that is for warm air that's already in the house to leave though the cracks around poorly fitted windows and doors. If you weatherize the house, you'll lose those leaks, and if the warm air can't escape, the cold air can't enter. Most ventilation ends when you weatherize. Interesting conundrum, isn't it?

They based the size of the hot-air flue on the square feet of connected indirect radiation. They allowed 1-1/2 square inches per square foot of radiation when they were heating with steam. They sized the cold air flue to be somewhere between two-thirds and three-quarters the size of the hot air flue.

If you have absolutely nothing else to go by, you can measure the length and width of the hot air flue to get an idea of what's happening. Multiply one by the other to get square inches. Then divide the total by 1.5. That will give you a good estimate of the square feet of radiation inside that duct. Another way to guesstimate is to look at the pipe size feeding the indirect radiator. The Dead Men would generally use a 1-1/4" pipe to feed up to 80 Square Feet E.D.R. of indirect radiation, and a 1-1/2" pipe to feed up to 100 Square Feet E.D. R. of indirect radiation.

Generally, the registers in the rooms are 25% greater in area than the flues that serve them. Again, there are no fans to move the air in this type of system. Everything works by natural convection. That means the air moves more quickly to the upper floors than it does to the lower floors because of the chimney effect of the taller, second- and third-floor flues. Typical air velocities are 1-1/2 feet per second to the first floor, 2-1/2 feet per second to the second floor, and 5 feet per second to the third floor. Notice how the air speeds up as it moves higher. Because of these differences in velocity, each flue served only one floor. And since the air moved more quickly to the upper floors, the Dead Men usually made these flues about 25% smaller than those serving the lower floors. They also used smaller registers on the upper floors. This can get tricky if all you're looking at is the register. And please don't try to equate any of this to a modern forced-air system. It's very different.

Because they used this system for ventilation, as well as for heating, they had to allow for more radiation. Their general rule of thumb in the old days was to take a heat loss of the space, using the Mills Rule (which I explained in the beginning of this chapter). The Mills Rule allowed for one Square Foot E.D.R for each 2 square foot of glass, each 20 square foot of cold wall, ceiling or floor, and each 200 cubic feet of room volume. They'd total these three things and come up with a radiation load, to which they'd add their standard Pick-up Factor for the pipe load. Once they had this figure, they'd add 25% more if the system was heated indirectly by steam. This allowed for the fresh air and for the limited convection currents in the rooms themselves. Consider how this can affect a replacement boiler size if you're not going to be bringing in fresh ventilation air. Nowadays, even wealthy homeowners often decide to abandon the ventilation side of their indirect systems so they can save some bucks on fuel. If they ask you to help with this, you can seal up the fresh air inlet and work only with the air in the basement. But you will have to find a way to get the upstairs air back down to the basement. Often, a louvered basement door is all it takes to make that work.

A quick story for you

Years ago, when I was doing consulting work, a company hired me to look at one of the homes owned by Doris Duke. She was one of the wealthiest women in the world at the time, and this turn-of-the-century mansion was in central New Jersey on a magnificent, 2,700-acre estate that made my head spin. The mansion was beautifully maintained and it had indirect radiation on several of the floors. It was just as I described above, but this one ran on hot water instead of steam, which made it a bit different, but not for the purpose of this tale.

Miss Duke's complaint was that the air rising from the bronze floor grates on the main level was too cold. I checked it with a digital thermometer and, sure enough, it was only about 75 degrees. That's much too cool to get the job done.

I went over all the mechanical equipment with the plumber, who, by the way, lived on the estate. Miss Duke had a member of every trade you can think of on her staff, which numbered more than 100.

Everything looked fine at first. Then we walked to the duct that fed warm air upstairs and I opened the access panel. The heat smacked me in the face. It was like opening the door of a pizza oven. The indirect radiator was roaring hot, but the air simply wasn't moving across it.

"I don't understand this," the plumber said. "Heat rises. The heat *should* be going upstairs, but it just won't move."

"Heat doesn't rise," I said. "Heat radiates. Warm air rises."

"Well, it won't rise here," he said.

"That's because there's no cold air to replace it. You can't move air without replacing it with more air." He just looked at me. "Where does this duct go?" I pointed to the left.

"I think it goes outside," the plumber said.

"Where does it come out outside?" I asked.

"I have no idea. I don't do outside. I only do inside."

We went outside anyway.

After looking around those gorgeous grounds for about ten minutes, neither of us could find the spot where the ductwork left the huge house. So we went back to the basement and measured in from the corner of the house to the spot where the ductwork met the outside wall.

Then we went outside again and measured in from the same corner. All I found there was topsoil and some newly planted groundcover.

I had a hunch, though, so I started to stomp around on the plants, and, sure enough, there came a hollow noise from the ground. I smiled at the plumber. "Go get us a shovel," I said.

The plumber found the gardener and came back with a long-handled shovel. We started to dig. When we got down about six inches we hit pay dirt. There was a 4' X 8' sheet of stiff plastic down there. The gardener had used it to cover a ground-level iron grate, which he thought was ugly. This grate, however, led to the below-ground vault that connected with the fresh-air duct for the indirect heating system. We removed the plastic and went back inside. Both of us smiled as we felt the now-very-warm air wafting beautifully from those ornate bronze floor grates.

In this case, the butler didn't do it. The gardener did.

Controlling the radiators

So those are the basic types of radiators that you'll see in a steam system. Knowing the E.D.R. output of all of them is important when you're replacing the boiler (see those sources I mentioned earlier for other types of radiators).

Now let's look at things, both old and new, that we can use to control the amount of heat leaving a radiator. This is good stuff because the more control you have over the temperature in the rooms, the greener your wallet will be.

Two-pipe steam

With two-pipe radiators, we have supply valves at the inlet to each radiator, which you can open or close to start or stop the flow of steam to any radiator. You can also throttle this supply valve, regulating the amount of heat that enters the radiator. It's all wonderfully simple, as long as the valves are working, and as long as you can get at them. Oh, and as long as the tenants are willing to do it. And if the tenants aren't willing to do it, there are other options.

I once visited one of the most prestigious addresses in the world. This landmarked apartment building is in New York City and it is home to some very wealthy and famous people. They all live with radiators that Dead Men installed in the late-1800s. When any of the rich and famous people are too warm, that person will call downstairs and a member of

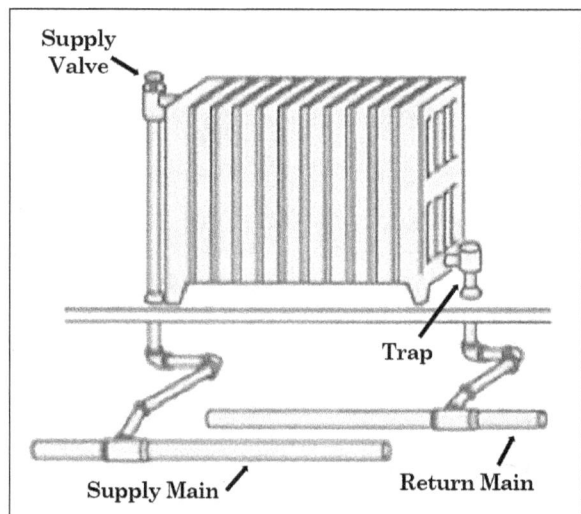

Two-pipe steam radiator

the building staff will ride the service elevator up to that person's apartment and close the supply valve on his or her radiators. The staff person then returns to the basement and waits for another call. When it comes, staff will return to the apartment to open the radiator valves for the now-too-cold tenant. This is what we call a two-legged zone valve, and it's an option only the rich and famous can afford. The rest of us have to do it ourselves.

Which brings us to automatic valves.

Pneumatic supply valve

This is a pneumatic supply valve. You use compressed air to open and close it. The valve is normally open and spring-loaded. When you bring compressed air to the bellows that sits on top of the valve, it will expand and close the valve. You'll see these in schools, hospitals, government buildings, and office buildings. The thermostats also work on compressed air, and you'll know you have one of these if you touch the thermostat and it makes a hissing sound (you're releasing air pressure when you do that). How well a pneumatic system works depends on how tight all the connections on the air lines are. Often, those connections are weak and they're leaking like a litter of puppies. The air compressor is running all the time, trying its best to maintain pressure, and that's wasting lots of electricity, which isn't very green.

TRV replacing pneumatic supply valve

So what you'll often see where there once were pneumatic valves is this.

This is a thermostatic radiator valve (or TRV for short). Can you see the abandoned air line in the background? Someone replaced the pneumatic valve with the TRV to regain control after the pneumatic system sprang too many leaks. The TRV comes in two parts – a normally open, spring-loaded valve, and a temperature-sensitive (and temperature-adjustable) controller. The controller is self-contained. It uses a chemical to sense the air temperature in the room. When the air gets warm, the chemical expands and pushes down on the spring-loaded valve, closing off the steam to the radiator. When it gets cold in the room, the controller responds by opening the valve. If you install these valves properly, they'll maintain the set temperature in the room within a couple of degrees.

The biggest problems I've seen with TRVs occur when someone places them inside a radiator enclosure. It's hot in there and the TRV sensor picks up on that right away, closing the radiator before the room comes up to the desired temperature. The best way around this problem is to get TRVs that have remote sensors. That way, the sensor can feel what's going on outside the enclosure, rather than inside the enclosure.

Another problem occurs when the installer snaps the controller onto the valve and doesn't take enough care in lining up the two. If the controller is cocked to either side (and this is easy to do if the installer is in a contortionist's position at the time), the valve stem won't seat properly on the controller's saddle and the radiator will overheat. Check on this if you have a radiator that has a TRV, but is still putting out too much heat.

When you have TRVs on the radiators, you still need something to control the boiler. If you're using an electric thermostat to do this, it should ideally be in the coldest room in the building, and there should not be any TRV-controlled radiators in that room. Set up this way, the boiler will try to satisfy the electric thermostat in that coldest room, and the TRVs will keep the other rooms from overheating.

Steam-trap maintenance is very important when you have TRVs on two-pipe radiators. We'll look more closely at how to green traps a little later, but I want to mention them here as well because TRVs can indirectly damage radiator steam traps under certain conditions. It works like this. Imagine you have a steam riser in an apartment building. It goes up 10 stories, feeding steam to 10 radiators. Each of those radiators has a TRV. You installed them to save fuel in the apartments by giving the tenants the ability to set individual temperatures. The steam climbs the riser and enters the radiators. The TRVs close tightly because the temperature in the rooms is at the setpoint.

With me so far? Good. Now let's say that a couple of the thermostatic radiator traps on the return side of this riser fail. They're now passing steam into the return line, which is never a good situation. You don't know this is happening because you're not checking the steam traps (too bad) and when a radiator trap goes bad, the person with the bad trap usually still has heat (ironic, isn't it?). There's a radiator with a TRV and a perfectly good trap nearby. The steam gives up its latent heat inside the radiator and the TRV and the trap on that radiator both close. Condensate forms inside the radiator, and, along with the condensate, a vacuum also forms. When the radiator trap opens, steam that's in the return line (coming through the nearby failed traps) enters the radiator with the closed TRV backwards. It does this because high pressure goes to low pressure – always. The return-line steam can now overheat the radiator with the closed TRV. And it can also damage that perfectly good trap through water hammer. Water and steam are not supposed to be going backwards through *any* steam trap.

So if you're installing new TRVs on an old steam system, be sure to check those steam traps. If you don't, you'll be wasting a lot of green.

One-pipe steam

You have to play this one differently. One-pipe steam doesn't have that second pipe to drain the condensate, so you can't throttle the supply valve. Well, actually you *can*, but if you do, you'll get knocking and banging and the air vent will squirt water all over the ceiling, so I'd rather you didn't throttle.

Back in the '70s, a few valve manufacturers tried to come up with a one-pipe-steam TRV that would take the place of the supply valve, but none of their attempts were successful. Then, they started looking at the challenge in a different way. The goal was to keep the individual radiators from overheating, and you could also do that by stopping the air from venting from the radiator once the room reached a certain temperature. If the air can't get out, the steam can't get in. So the focus shifted away from the supply valve on that one-pipe radiator, and toward its air vent.

So let's put the TRV *between* the air vent and the radiator.

Okay, we'll start with a cold radiator that's filled with air. Steam arrives at the supply valve and starts working its way to the top of the radiator, pushing air ahead of itself. The air moves through the normally open body of the TRV and leaves the system through the normally open air vent.

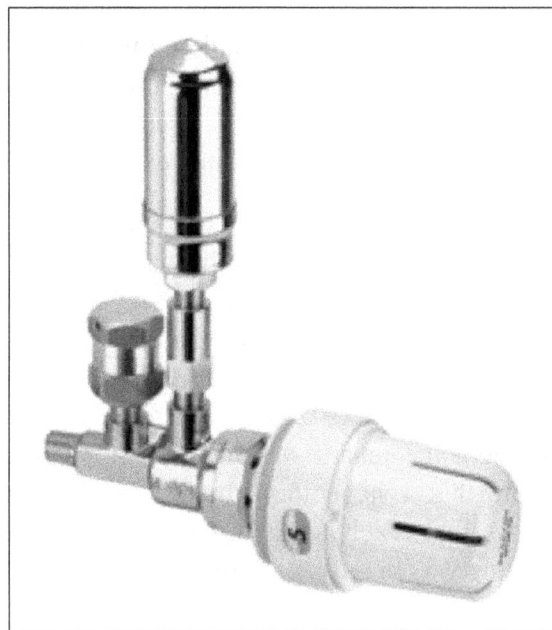

One-pipe steam TRV

Oh, and while I'm thinking of it, it's always better to use a straight-shank air vent with a TRV because condensate will drain much better from these than it will from the angle-shank air vent that's probably on the radiator before you add the TRV. Gravity likes a straight shot downward.

The steam is heating the radiator now, and the radiator is heating the air in the room. That warmed air is bathing the TRV's controller and the chemical inside that controller is expanding. Before long, it will create enough pressure to close the normally open valve that's between the radiator and the air vent. As soon as that happens, air stops leaving the radiator. And if the air can't get out, the steam can't get in. The result is that the room doesn't overheat, which is nice and very green indeed.

Now the next thing that happens is that the steam in the radiator is going to condense, and since the TRV's valve body is closed, no air can enter the radiator through the air vent, even if the air vent cools and reopens. This was a major problem with the first generation of these one-pipe-steam TRVs. They closed; vacuum formed, and the radiator kept heating – even though the TRV was satisfied. Manufacturers of these devices solved the problem by adding a vacuum breaker to the TRV's valve body. Now when the vacuum forms, the vacuum breaker senses it and allows air back into the radiator. That stops the steam's advance and keeps the radiator from overheating. Beautifully simple!

This type of TRV often winds up in buildings where there is a store on the first floor and apartments above. The thermostat for the whole building is probably inside the store, and the door to the store keeps opening and closing all day long, calling on the heat. The people in the apartments are roasting so they open their windows, and out flows the fuel dollars.

By adding TRVs to the rooms that are not near the electrical thermostat, you'll prevent overheating in those spaces and save fuel. That's an important point. **Don't** use a TRV in the room that has the electrical thermostat. And ideally, that room should be the coldest room in the building.

Got it? Great! Let's take another look at an oddball.

Two-pipe, air-vent radiators

A story for you. The building was on Fifth Avenue in midtown-Manhattan, and that's one of the reasons why this one has stayed with me for so long. It was so stately and classy. It stood there with all the modern skyscrapers, looking very much like an aristocratic old man at a gathering of newly rich, young executives.

I was with a guy who had gotten his professional engineering license around the time I was born. He had brought me along for a couple of reasons. First, he thought I'd find the place interesting, which I did. Second, because he wasn't positive about what to do to fix the problem the tenant was having. And third, he knew that I have this library of engineering books that reach back to a time when central heating was brand-new. Comes in handy.

He told me that the tenant no longer trusted the engineering firm they had hired to write the specs for the renovation of the offices they occupied. None of their radiators would get hot. They used to get hot, but they didn't anymore. That was the problem. They hired this engineer because he was older than most of the people in New York City. They figured he'd know more about this stuff than most, which he did.

Two-pipe, air-vent radiator

The project occupied just two floors of the grand old 15-story building. The tenant bought the space in this building, had it gutted, and then set out to make it look modern. The steam system that heated their two floors (along with the rest of the building) however, dated to Ragtime. Folks who buy space in steam-heated buildings often think that if they modernize their space the old heating system will watch, and nod, and also become modern. The new folks change the radiators, hide the pipes, and wonder why, with all the money they've spent, things aren't going their way. This shouldn't be happening to them, but it is, and the reason why is because the rest of the building didn't go along with their modernization.

In this case, the new tenant (these folks who had bought just two of the 15 floors) tried to fiddle the old system into the 21st Century. The consulting engineer, eager to please his client, decided to modernize just the portion that concerned the tenant by replacing the old cast-iron radiators with sleek, European-style, panel radiators made of steel, with very narrow internal passages (never a great idea with steam). The consultant also added thermostatic radiator traps, and thermostatic radiator valves to each radiator. He explained to his client that these modern additions would give them total control over the heat, but when autumn arrived, the only available heat came from the computers.

So one guy pointed a finger at another guy, who pointed at a third guy, who looked around for a fourth guy, who pointed back at the first guy. And you know the rest of that story. Which is why the oldest engineer in the City of New York and I were there that day.

Which brings me to the elephant. There were these blind men who came upon an elephant for the first time. One blind man touches the elephant's leg and tells the others that an elephant is like a tree trunk. But another blind man touches the elephant's tail and decides that the first blind man is an idiot. An elephant doesn't look like a tree trunk; it looks like a snake. And then the other blind guys step up. They each grab a different part of the elephant, and each comes up with his own conclusion as to the true appearance of an elephant. None of them take the time to explore the whole elephant, and that's what's causing the confusion and the disagreement.

The consulting engineer on this project decided to touch just two stories of this old elephant of a building. He saw antique radiators without steam traps, and declared that they must add traps because all steam systems of the two-pipe variety require steam traps (not true). I imagine the consulting engineer also decided that the long-dead design engineer was a know-nothing fool. Who would design a two-pipe steam system without steam traps? Indeed!

The thing that should have nagged at the consulting engineer, though, was that this wonderful old building had heated splendidly for more than 100 years. And it did that without benefit of steam traps at the radiators. It's hard to argue with success, but the consulting engineer didn't consider that.

So the oldest engineer in the State of New York and I walked upstairs and took a look at the heating system in the neighbor's space. This was how it used to be in the new tenant's space. One look was all it took to tell us what we had here. The tip-off was the size of the supply and return risers. On most two-pipe steam systems, you'll find big supply risers (because low-pressure steam is big), and small return risers (because condensate is small). There might be, say, a 1-1/4-inch supply and a half-inch return. Here, we had a 1-1/2-inch supply and a 1-1/4-inch return at each radiator. There was also a one-pipe-steam air vent on the return side of each of the original radiators. There were no air vents on the modern, steel-panel radiators, of course. They had thermostatic radiator traps and thermostatic radiator valves. Modern stuff. None of it worked, but it looked marvelous.

The oldest engineer in the United States of America smiled at me and I smiled back. We knew that what we had here was a two-pipe, air-vent system. That's why the new tenants with their newly modernized portion of the system had no heat.

There was a time in American Heating History when they didn't use thermostatic radiator traps. They didn't use these devices because no one had yet thought to invent them. You cannot install what has not been invented. This lack of traps, however, didn't stop the Dead Men from installing steam heat. They just put in these two-pipe, air-vent systems, which look remarkably like two-pipe, direct-return hot-water systems. The steam leaves the boiler and heads up into the building. It favors the supply lines, of course, because these are usually larger than the return lines by at least one size. Steam follows the path of least resistance. The Dead Men put an angle valve on each side of every radiator so that tenants could shut off the heat if it got too warm. The air left the radiators through the one-pipe-steam air vents. The steam heated the radiators, and the condensate dribbled down the return lines. The steam flowed into the return lines along with the condensate because there was nothing there to stop it. After a while, there was steam everywhere, and that was perfectly normal for this system. And since there were so few moving parts, these systems lasted for as long as the building stood.

But then the consulting engineer came along and decided that he could put a party hat on the elephant, while leaving the rest of its body naked. He added the traps to a system that's not supposed to have traps because there's supposed to be steam in the returns (the new thermostatic radiator traps shut, and stayed shut). He got rid of the air vents on the radiators (even though they were supposed to be there), so the air had no way out of the system. And he added thermostatic radiator valves, which then stayed opened because the rooms were as cold as could be. Steam couldn't enter the radiators because air couldn't leave. And if steam did enter the radiators, the condensate would have no way of leaving because the steam in the return kept the traps closed.

The oldest engineer in the Western Hemisphere and I told the tenant that they were either going to have to return to the 19th Century, as far as their heating system was concerned, or convince the rest of the tenants in the building to move boldly along with them into the 21st Century.

And this being New York City, I'll leave you to imagine what the other tenants had to say about that.

Balancing the load

Let's briefly focus again on that the definition of Square Foot of E.D.R., the term we use to measure the output of radiators because it plays a large part in system balance. For steam, one Square Foot E.D.R. will emit 240 Btuh when there is 70° air on the outside of the radiator and 215° steam on the inside of the radiator. That's what we want to have happen on the coldest day of the year. On milder days, the radiators are oversized and it's normal for them to get hot just partway across. That's balance.

The trouble is this usually doesn't happen. Some radiators get too hot while others stay cool. This wastes money because people open windows when they're too hot, and complain when they're too cold, causing you to run the burner longer for them (as the others are sweating). As we move along into the other chapters of this book, I'll talk to you a lot more about balancing steam heat so that everyone is happy. For now, though, let me just touch on the elements that make for balanced steam systems:

The elements of steam system balance

First, there's **water quality**. If the water is contaminated with dirt, or if its pH is wrong, it won't boil properly and you'll have wet steam. The steam will give up its latent heat to the liquid water that it's carrying along and stop moving toward the radiators. If the steam doesn't go where we want it to go, the system will not be in balance.

Next, there's **air venting**, and we'll spend a whole chapter on this one. If the air can't get out, the steam can't get in, so venting is critical to steam system balance.

Steam traps play a *huge* part in system balance, and these deserve their own chapter, which is coming up. I often compare traps in a steam system to balancing valves in a hot-water system. Traps set up the points of high pressure and low pressure, as well as high temperature and, in the case of thermostatic radiator traps, lower temperature. When traps go bad, balance goes out the window. So do the Btus.

Size matters. And when it comes to **pipe size** it *really* matters. At the low pressures we use for steam heating, a certain amount of steam will flow through a pipe at a definite pressure drop (typically 1 ounce of loss per 100 feet of travel). If the pipe is the wrong size, it won't deliver the goods at the right pressure. The result is an unbalanced system. Steam, like water, follows the path of least resistance.

And finally, there's **pipe insulation**. When steam turns to water, it stops being steam. And it stops moving. Some folks will be hot; others will be cold. You'll be wasting fuel, and that's not green.

There's much more on all of these elements coming up. And since I just mentioned pressure, let's take a look at that. Pressure has a *lot* to do with how much green you'll be spending on that steam system.

CHAPTER 3

GREENING THE PRESSURE

Low pressure goes fast!

In the mid-1980s, when I was working for the manufacturers' representative, I was spending a lot of time in New York City, looking at old steam systems and trying to make sense of what I was reading in the books that the Dead Men had written. New York is a heating museum.

New York City's Department of Housing Preservation and Development was doing a lot of work with the steam systems in the old apartment buildings back in those days, and there was a group of enthusiastic people working either for H.P.D., or as consultants to them. Energy conservation was a new concept and we were all learning together. One of the people in that group was a young Professional Engineer named Frank Gerety. Frank's passion was steam heating, and in 1987, he prepared a booklet for H.P.D, which he titled, *How to Get the Best from One-Pipe Steam – A Practical Manual for Heating System and Boiler Optimization, Maintenance, Loss Prevention, Replacement, and Installation.* That booklet (now out of print) guided most of us who were working with steam at the time, and I would spend hours at Frank's knee, listening to him talk of the Dead Men, and asking all the questions I could think to ask.

I was in his office one day when he said something to me about steam heating that caused all the confusion in my mind to clear. It was as if he flipped a switch. I was telling him about a problem steam system that I had visited with a contractor, and I mentioned that we had turned up the steam pressure in an attempt to get the steam to move more quickly from the boiler to the radiators. Frank laughed and said, "Dan, if you want to steam to move *fast*, you don't *raise* the pressure; you *lower* it."

That made absolutely no sense to me. I figured that high pressure meant speed. Push harder and things go faster, right? So I asked more questions and Frank grabbed an engineering textbook and turned to the page that had this drawing of what's called a Moody Friction Flow chart. Being a non-engineer, this chart looked like a bowl of linguini to me. There were lines running everywhere. Frank looked at me and laughed again. And then he explained it all in a way that I could grasp. Frank was an excellent explainer. He painted beautiful word pictures, and this is how he got it through my thick skull.

When a boiler turns water to steam, it causes the water to suddenly expand 1,700 times, and that's at 0-psi pressure. This massive expansion drives the steam out of the boiler and into the pipes. The steam is moving very quickly at this point because its volume is so large. A gas that isn't under much pressure will expand as much as it can. If you add pressure to that gas, its volume will decrease, so when you raise the pressure on your steam boiler, the steam doesn't go *faster*; it goes *slower* because of its volume.

Seems crazy, doesn't it? That's what I thought at first. It goes against logic. You would think that if you raised the pressure at the boiler you'd be pushing the steam harder, so it should get to your radiators quicker, right?

It doesn't. And here's why.

Take a look at the nameplate on any boiler and you'll find a rating in **Btuh**, which is short for **B**ritish **t**hermal **u**nits per **h**our. Consider that term for a moment. One Btu will raise one pound of water one degree on the Fahrenheit scale. Btuh tells us how much water we can heat in an hour. These are finite terms – pounds of water, and hours on the clock. You can't get more heat out of that boiler than the fire puts in. It's just like an ATM. You can't take out more money than you deposit. Raising the steam pressure on that boiler won't increase the amount of heat that comes from the fire each hour. You simply can't get more out than you put in, and what you put in is what the nameplate on the boiler states.

So consider the term Btuh. The "h" stands for "hour." In one hour, that boiler will produce a very definite amount of steam. It won't do this in 59 minutes, or in 61 minutes. It will do it in *exactly* one hour. That's very important, as you'll see.

Okay, now think about the pipes through which the steam will be traveling. The size of those pipes is fixed, isn't it? I mean no one is going to show up and change the size of the pipes when the boiler starts to make steam, are they? Of course not. The steam will enter those pipes and it will flow toward the air vents. And as it does, it will lose pressure along the way due to friction (caused by the steam rubbing against the insides of the pipes). The steam will also lose pressure as it condenses.

So we have a locked-in amount of steam (because of the size of the burner) traveling within pipes that have a fixed amount of internal space. With me so far? Great!

Okay, let's consider the steam, which is a gas. When you compress a gas it takes up less space, doesn't it? Sure it does. You know this from everyday life. Just think about helium for a moment. You go to the store to buy a bunch of balloons for a birthday party. When you get there, you see that other people are also having birthdays. The ceiling at the party store is crowded with bouquets of balloons, all waiting to be picked up by happy people.

Take a moment to notice how much space those balloons are taking up as they press against the ceiling in the party store. Now look at that tall helium canister that the balloonologist (good word, eh?) is using to fill all those balloons. See how small it is compared to the volume of all those balloons? That one helium tank at the party store will fill about 600 balloons, each 11 inches in diameter. But how do they get that much helium inside that steel tank?

They pressurize it, of course (which is why the tank is made of steel), and when you open that tank the helium comes roaring out and expands like mad. Gases do that when you let them lose. They expand and contract in relation to the pressure that you apply to them.

Okay, back to our steam system. The Dead Men designed their systems to have steam moving as quickly as 60 miles per hour, and that's based on it leaving the boiler at a pressure of just a few ounces per square inch (which is what we have when the boiler first starts making steam). Now one pound of liquid water will fill a single pint glass. Turn that pint of water into a gas (steam) at zero pounds per square inch and it will fill 27 cubic feet. That's one cubic yard, three feet on each side! Raise the boiler pressure to 10 psi and that same pound of steam will take up just 16 cubic feet. It's much smaller now. The higher you raise the pressure on the boiler, the more compressed the steam will become. It's like the helium in the tank at the party store.

Hold that thought for a moment.

Now imagine in your mind's eye 1,000 Smart Cars. See them in a rainbow of colors and put a driver inside each of those little cars. Let's call each driver Mr. Btu.

Now imagine a one-lane road that's a mile long. It can be a country road or a highway; it's your choice, but what you're going to do in your imagination is move those 1,000 Smart Cars down that mile-long road in precisely one hour. That's the goal – to move 1,000 Mr. Btus per hour. You have to do it in such a way that the last car crosses the finish line when the stopwatch hits the one-hour mark. You can't start a moment sooner, and you can't finish a second later. Can you see it? Good. This is what we mean when we talk about Btuh (British thermal units per **hour**). If I say 1,000 Btu**h**, I'm not talking about 1,000 Btus in 59 minutes, right? Nope, I'm talking about that much heat in one *hour*.

So when comparing the movement of steam to the movement of Smart Cars (something you do every day, I'm sure), please keep time in mind.

Okay, what else? Oh, did I mention that those Smart Cars are under 10-psi pressure? Well, they are. Add that to your imagination. You've got the Smart Cars moving down the road (or through a steam main) and they're all under 10 pounds per square inch pressure, which is why they're so small.

Okay, time to get rid of the Smart Cars. We'll release the pressure on them, and when we do that, they'll expand like a gas. In fact, they'll magically transform themselves into big yellow school buses (work with me on this).

Now, bring the buses around to the starting point. Who's driving? The 1,000 Mr. Btus – the same guys who were driving the Smart Cars. It's the exact same amount of heat, just moving in larger vehicles. Your job is to be in charge of the team that's going to move those 1,000 big yellow school buses down the same road that the Smart Cars just traveled. And you have exactly one hour to do this. You can't take a second more or a moment less to get the job done; you have to do it in *exactly* one hour.

Ready for the big question? Which vehicles have to travel *faster* to get the job done – the Smart Cars, or the big yellow school buses? The drivers are the same. The road is the same. The time is the same. The only thing we've changed is the size of the vehicles. So, which line of vehicles has to go faster to cover the distance in exactly one hour?

The school buses, right? Of course, they do! They *have* to go faster because they're bigger. Just think of how much more space a line of 1,000 school buses would take up, compared to a line of 1,000 Smart Cars. If you have to get those two lines down that mile-long road in *exactly* one hour, the buses will naturally have to go faster than the Smart Cars. Make sense?

That's why low-pressure steam moves faster than high-pressure steam. It's bigger. Think about it:

- ✓ If the quantity of steam the boiler produces in one hour is fixed (which it is).

- ✓ And if the pipes will always have the same amount of internal space (which they will).

- ✓ And if that steam takes up more space when it's at low pressure than it does when it's at high pressure (which it does!).

- ✓ Then we can say with absolute certainty that our given amount of steam, over the course of that one hour, within those pipes of a fixed size will *definitely* move faster at low pressure than it will at high pressure. It has to. At low pressure, the steam is as big as a school bus, and it's turbocharged with latent heat.

And that's how Frank Gerety, Professional Engineer, got the concept of steam velocity through the thick skull of his student. I hope that the student has done his teacher justice.

The Dead Men who designed steam systems designed the pipes to deliver steam to the radiators at *very* low pressure. The lower the pressure, the faster the steam will travel (and the greener the system will be). If the steam isn't getting to the radiators, it's not because of a lack of steam pressure. Something else is going on.

Danny Pressuretrol

John Chapman traveled the Midwest during the early-1800s, wearing his cooking pot on his head. He owned a bunch of nurseries in Ohio, Pennsylvania, Kentucky, Illinois, and Indiana where he grew apple trees. He gave most of them away as he traveled. Folks called him Johnny Appleseed.

After my days with Frank Gerety, I became Danny Pressuretrol. I traveled the basements of the New York/Metro area with local contractors, showing them how to crank down the pressure to make the steam go faster. I was able to do this because I had a counterclockwise screwdriver. Here's a picture of it.

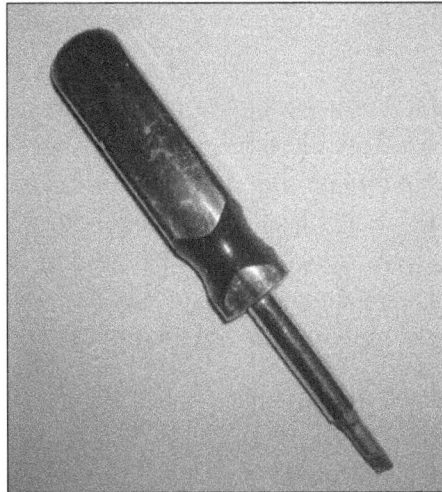

Counterclockwise screwdriver

This remarkable, and highly unusual, tool was able to turn the screw on a steam system's pressuretrol in a counterclockwise direction, which forces that system to run at a lower pressure. Most contractors I met owned *clockwise* screwdrivers, which could turn only in *that* direction when applied to a presssuretrol's adjustment screw. Theirs looked like this.

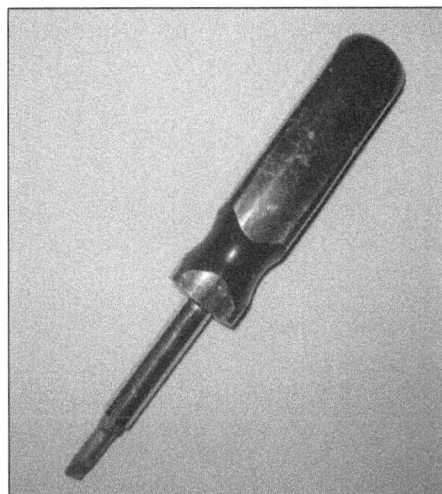

Clockwise screwdriver

These contractors had been turning up the pressure for so many years that their tools had picked up the habit. They marveled at my screwdriver, which could go the other way. "Crank it down!" I'd shout, and they'd give it a try.

One contractor told me of a job where he lowered the steam pressure so that air could escape (much more on this phenomenon coming up). The building heated faster and the gas bill went down by such an impressive amount that the local utility sent an inspector to see what had changed. They thought someone had bypassed the gas meter. Crank it down! Crank it down! Crank it down!

But not always.

When I first started in business, I was doing a lot of consulting. I'm not an engineer, but I did have that screwdriver and all those old books written by the Dead Men. One day, the local power utility hired me to help figure out why this steam unit heater that hung from the ceiling of one of their warehouses wasn't heating as it should. I looked at it and saw that it was a two-pipe unit heater, and that the steam traveled down to the unit from an overhead steam main. The condensate left the unit through a bucket trap and had to travel upward to the return main, which was slightly below the steam main. So they had to lift condensate.

"What's the pressure?" I asked the guy.

"It's two psi," he said.

"But you have to lift the condensate all the way up there," I said. One pound of steam pressure will lift condensate about two feet. You have two pounds of pressure so that's good for only about four feet. Check out that lift up there." He looked up. "It's more than four feet, right?"

"It sure is," he said.

"Why do you have the pressure so low?" I asked.

"Well, last year, we sent our engineers to this guy's seminar. He's supposed to be some sort of steam guru and he kept saying, 'Crank it down! Crank it down!' So we did."

"That was me," I said.

"That was you?"

"That was me."

"So what do we do now?"

"Crank it *up*!" I said.

That solved the problem.

There are exceptions to some rules, but with steam heating, unless you're lifting condensate, you always want to apply that special counterclockwise screwdriver and take Danny Pressuretrol's advice.

Crank it down!

Two-psi should do it

You should be able to heat nearly every steam-heated building built in the United States of America after 1899 with 2-psi pressure or less. If you can't, the problem isn't the pressure you're using. Trapped air is probably causing the problem.

During the mid- to late-1800s, contractors were using steam pressure to compete with each other. One contractor would figure a job at a safe, low pressure, say, 2-psi at the boiler. Another contractor would figure that same job at a much higher pressure (sometimes as high as 60-psi). This guy knew that as the pressure got higher, the steam compressed into a tighter bundle, so he figured he could use smaller pipes and smaller radiators. The problem was that both guys had to start out each steam cycle at 0-psi pressure because that's the way steam heating works – you begin with a boiler that's off. Low-pressure steam (which is what both guys had at the start of each cycle) travels *much* faster than high-pressure steam (you know that already). The installer who sized his pipes for 60-psi pressure had steam moving through his pipes like crazed NASCAR drivers every time the boiler started. That high-speed steam zoomed out of the boiler at such an insane speed that it pulled the water with it. The water banked up the sides of the pipes and crashed into every fitting and valve it met along the way. The water wound up in the system and the boiler kept running because there were no low-water cutoffs in those days. Under these conditions, a boiler could burn up, or even explode, and they did.

And that brings me to the Carbon Club.

On December 18 and 19, 1899, they held a special meeting at the Murray Hill Hotel in New York City. The managers of the many American boiler companies made up the membership of this recently formed group. During that meeting of the Carbon Club, the Committee on Boiler Ratings recommended a standard for boiler pressure, which the Club accepted. It became the standard for house heating. The standard applied to both steam- and hot-water boilers, and its intent was to remove "trade perplexities" in both sizing and pricing. They were leveling the playing field.

From that day on, all boilers would be rated on a proportion of 100 for steam and 165 for hot water, with steam at 2-psi pressure at the boiler, and water at 180° Fahrenheit at the boiler. You can see this today in the values we assign to the term E.D.R. The rating also considered all the mains, returns and risers as heating surfaces, and stated that an installer must include these surfaces in his sizing, so as to avoid an undersized boiler. This established the Pick-up Factor for the first time.

To make it all work, they came up with standard pipe-sizing methods, which, for steam, allowed for a pressure drop of about one ounce for every 100 feet of travel. Those are the pipes that are installed in the building that you are trying to make greener. The Dead Men set the required pressure for that job, and it's not more than 2-psi at the boiler. Trust me. If you can't heat a steam-heated building with 2-psi pressure or less, the problem is *not* the pressure. It's something else, and most likely it's trapped air.

Consider the Empire State Building, which opened for business on May 1, 1931. There are 50 miles of steam pipes within the Empire State Building and about 6,500 radiators. On the coldest day of the year, the Empire State Building heats with less than 3-psi pressure. They do have a vacuum system in the Empire State, however, and that does increase the pressure differential across the system, so let's spend a moment with that.

A bit about vacuum systems

Steam systems that make vacuum with pumps have been around for more than 100 years. In 1899, D.F. Morgan wrote an article in *The Metal Worker* magazine, suggesting that designers skip the pump and let the steam make its own vacuum by expanding when it formed, and contracting when it turned back into condensate. He wanted to take advantage of that 1,700:1 ratio between water and steam. The trick was to make all the pipe joints tight so that air couldn't get back into the system once the steam had shoved it out, and to use special air vents that had check valves at the outlets. When steam turned into condensate and shrank, you'd wind up with a naturally formed vacuum and water would then boil at a lower temperature. This took advantage of the lower Btu input of a coal pile that was burning down later in the day.

These systems worked well, but only if they burned coal. When we began to use oil and natural gas instead of coal as a fuel, these systems became automatic. A thermostat started and stopped the burner, and when the fire stopped, there wasn't as much residual heat left in the boiler (compared to coal). The vacuum formed quickly and that was the problem. Not all the air escaped from the system on the first firing cycle, and when the vacuum formed, that air expanded tremendously. It moved out of the radiators and down into the pipes. The expanded air stopped the steam from reaching up to where the people were. So when the fuels became automatic, the vacuum vents on the radiators and near the ends of the main became a problem, so the Dead Men removed them. Or at least they were supposed to remove them. If you have an oil- or gas-fired system and it has this type of vacuum vent, replace those vents with standard air vents. Your system will balance better.

There were some other systems that used mercury-filled devices to make vacuum. Trane was one company that made such a device. It allowed the expanding steam to push the air through the pipes and the radiators, and then into a tube that dipped into a pot of mercury. The air would bubble up through the mercury, and when the steam collapsed back into water, air couldn't reenter the system through the mercury pot, which acted as a liquid check valve. If you ever run across one of these devices, please be careful with it. Mercury is no longer something we should be rolling around in our hands, as we did when I was a kid. It's definitely *not* green.

Vacuum systems that used pumps to make the vacuum have been around even longer. You'll see them on big jobs, such as the Empire State Building, and the advantage of having one is that the pump increases the differential pressure across the system by lowering the return-side pressure. Typically, a vacuum pump runs between 3- and 8-inches of mercury vacuum. It tries to maintain an average of 5.5 inches of mercury. Two inches of mercury vacuum is equal to one-psi of positive pressure, so when you add the positive boiler pressure to the negative vacuum pressure, you get the total system pressure differential. If the folks in the Empire State

Building run 3-psi pressure on a frigid day in February, and the vacuum pump is maintaining about 3-psi negative pressure (or about six-inches of mercury vacuum) the system will have a 6-psi pressure differential from supply to return.

Now think about what happens if the vacuum pump fails because the steam traps have failed and are passing super-hot condensate to the pump, causing it to cavitate. The people in charge of the building may decide not to have the pump repaired because it costs too much to do that. That's a real bad decision because if the pump's not doing its job, all of their pipes, valves and fittings will be undersized. The pump isn't pulling the air through the pipes, so the steam now has to do all the work of shoving that air ahead of itself, and the pipes are resisting. So the steam gets tired and slows to a crawl. Some of the tenants now have to wait a long time for their heat. And while they're waiting (and complaining), other tenants are getting too much heat. These people are opening the windows. The person in charge of the boiler will raise the pressure, trying to make all the tenants happy, but that just burns more fuel and wastes more money.

Once it's a vacuum system, it's *always* a vacuum system because the Dead Men sized every pipe, valve, fitting, and steam trap in that building for the pressure-to-vacuum differential. Without the pump, nothing works as it should. And the reason why the pump failed is probably because the steam traps failed. We'll get to that in just a little while, but first, let's talk some more about the glories of low pressure.

Why you can get by with low pressure

Latent heat does the trick. Think about it. Water can be either a liquid or a gas at 212° F, but which is *hotter*, the liquid or the gas? Both register 212° F on the thermometer. Which is hotter?

Good question, isn't it? Can you see how temperature is a lousy way to measure heat, even thought we use it all the time? (Whew! It must be 98 in the shade!) Measuring heat with temperature can really mess us up when it comes to steam heating. It's far better to measure heat with British Thermal Units.

The Btu was the brainchild of an Englishman named, Thomas Tredgold. In 1836, he wrote a book he titled, *The Principles of Warming and Ventilating Public Buildings*. In his book he writes, "In order to compare the effects of different kinds of fuel, some convenient measure of effect should be adopted; not only for the purpose of lessening the trouble of calculation, but also to render it more clear and intelligible. I shall therefore, without regarding the measuring of effect employed by others, adopt one of my own, which I have found useful in this and other inquiries of similar nature. I take as the measure of the effect of a fuel, the quantity of pounds avoirdupois, which will raise the temperature of a cubic foot of water one degree of Fahrenheit's thermometer."

There you have it, the first Btu, and probably not what you expected because Mr. Tredgold is talking about the amount of heat needed to raise *one cubic foot* of water one degree on the

Fahrenheit scale. Today's Btu measures the amount of heat needed to raise *one pound* of water one degree Fahrenheit, not one cubic foot. Those who followed Tredgold were able to do this because he was dead and not around to argue about it. So there.

There's nothing sacred about the Btu. It's just an agreement between folks who do heating for a living. When we all agree on the meaning of a term there can be peace in the valley. That's all it means.

So, to get one pound of water (that's about a pint) from 32° F to 212° F, we need 180 Btus (one Btu raises one pound one degree). We now have liquid water at 212° F. It's not steam. To get it to change state, to make those water molecules so excited that they leap off the surface of the liquid water and head for the radiators, we need to add an additional (and astonishing) 970 Btus per pound of water. That's more than *five times the amount of heat* it took to get us from 32° F, and at this point, the steam is at 0-psi pressure. This is the stuff that heats the building, and it's there at no pressure at all.

And that brings us to this next question:

What moves the steam?

It moves itself! No pumps required.

When steam forms, it goes off like microwave popcorn. I mentioned earlier that the expansion ratio is 1,700:1. Imagine a pint glass, one that you'd use to hold a big, cold draft beer. It's tapered, right? It's wider at the top than it is at the bottom, but if we measure in the middle of the glass we'll come up with about 2-1/2" in diameter. Think of that pint glass as a pipe that we're going to fill with steam. How tall is it? About six inches, right? It's filled to the brim with water (the beer's for when we're done working) and we're going to turn that water into steam.

Okay, how much pipe will that one glass of water fill with steam?

Well, I figure it would be 1,700 times the height of the pint glass, which is about 850 feet. That's about 85 stories straight up in a building, and that's at 0-psi pressure. This is why the quality of the steam, the insulation on the pipes, and the air vents are so important. It doesn't take much water to fill a steam system with steam, and the steam will travel all by itself. We don't need pumps to move steam. All we need is a bit of pressure to overcome the very low pressure drop presented by the system piping, and properly positioned air vents. This is one of the delightfully green things about steam heating. You don't need to pump it. That saves electrical energy.

The steam will do the opposite when it gives up its latent heat and collapses back into water. It will shrink to 1,700[th] of its volume, and as it does, it will draw the air back into the piping and radiators. It works like a bellows. On the next cycle, the expanding steam will shove the air out again, and so it goes, over and over again.

If you can't heat the building on low pressure, it's probably because the air can't get out. If you're having a problem with distribution, try this: Remove the main vent, which should be near the end of the main. Set the pressuretrol to cut in at ½-psi pressure with a differential of 1 psi. Start the boiler and see how long it takes for the steam to travel from the boiler to the point where you've removed the vent (be careful not to get burned). If the steam gets there at that low a pressure, then that's all the pressure you'll ever need. We'll talk more about this when we get to the chapter on Greening the Venting.

And since I just mentioned the pressuretrol, let's look at how to set one.

How to set a pressuretrol

The Pressuretrol controls the pressure. Great name, don't you think? Gets right to the point. I looked at these things for years and couldn't figure out how to set them. Most have a sliding scale on the outside, with a note letting us know that this is the Cut-In setting. And when you take the cover off the pressuretrol, you'll see a small dial on the inside that carries the label, "Diff," which is short for "Differential." But how do you know what to do next?

Let's start, once again, with that definition of EDR (yes, it's *that* important). One Square Foot of Equivalent Direct Radiation will emit 240 Btuh when there's 215° steam inside the radiator and 70° air outside the radiator. Since 215° steam is steam at about 1-psi, pressure, we know by definition of the term EDR that we're looking for 1-psi at radiator on the coldest day of the year.

So what do you need at the boiler to make that happen?

I spent a lot of time at the tradeshows when I was young, asking all the boiler guys how to set the pressuretrol. They told me to set it for whatever it takes, which wasn't very helpful. One manufacturer suggested that I contact Honeywell, since they're the folks who make most of the pressuretrols, so I went over to the Honeywell guy over on the other side of the tradeshow and did just that. He promised to send me the instructions and I went home and anxiously waited for the mail.

The instructions arrived a few days later, as promised, and I tore open the envelope in great anticipation. This is what it said:

"Set the cut-in scale to the desired pressure."

How nice. Whatever I desire. Gracious as all get-out, but not very useful. So it was back to me. What about the cut-out setting? I read on.

"Set the differential wheel to the number of pounds that the pressure should rise above the cut-in setting."

Isn't that delightful? The cut-in setting is whatever we desire, and the cut-out setting is whatever it should be. Wonderfully . . . vague.

So I went back to the old books and studied some more. I learned about the Carbon Club, and how, in 1899, they had established that all steam-heating systems for house-heating would work with 2-psi pressure or less. I studied air vents, and learned that the pressure within any steam system has to fall after it rises so that the floats inside the air vents can fall and allow for more venting after the first cycle. If the vents don't reopen, the rest of the air can't get out of the radiators. And where there is air, steam will not go.

So you set the pressuretrol for ½-psi cut-in, with a 1-psi pressure differential. This will have the system operating between ½-psi pressure and 1-1/2-psi pressure whenever the thermostat is calling. I'd say set it lower but that's as low as the standard pressuretrol goes.

The Carbon Club gave us pipe-sizing charts that would allow the steam to move through the pipes while losing just one ounce of pressure for every 100 feet of travel. That's not much. If you start out with 2-psi pressure at the boiler, 1,600 feet later, you'll still have 1-psi pressure available. That's simplifying things because we have to take into consideration the additional pressure drop of pipe fittings and valves, which offer more resistance than straight pipe, but I think you get where I'm heading. When in doubt, crank it down.

There's another type of pressuretrol (just to confuse things) and with this one, you have to subtract rather than add. You'll know you have one of these because the front of the control will have two scales. One will read as "Main" and the other will read as "Differential." There's probably also a note on the control that tells you, "Differential is Subtractive." Here's the deal with this one. "Main" is the cut-out pressure, the pressure that will shut off the burner and stop the steam from moving. That's the stop point. From this, you *subtract* the Differential. So if you wanted the system to operate between ½-psi pressure as the burner cut-in point, and 1-1/2-psi pressure as the burner cut-out point, you'd set the Main for 1-1/2-psi pressure, and the Differential for 1-psi pressure.

Get it? 1-1/2 minus 1 equals ½.

Why do the manufacturers offer two types of pressuretrols?

Because they can.

Crank it down!

For space heating, always set your pressuretrol as low as possible. If it's a standard pressuretrol, the type that comes with a packaged steam boiler, that would be a ½-psi cut-in with a 1-psi differential. I find that these standard units aren't that accurate, but do the best you can. Crank it down.

And if you're concerned about the burner short-cycling when you're operating the boiler at that low a pressure, don't be. If the burner short-cycles, it's probably a venting issue. The air's not getting out of the steam's way quickly enough. We'll talk a lot about this when we get to the chapter on Greening the Venting, but for now, just know that you can move with a pinhole what you can't move with a ton of pressure. Raising the pressure in a heating system may *seem* to cure burner short-cycling, but all you're doing is wasting fuel by forcing the burner to run longer. The folks upstairs are opening the windows. Go outside and see for yourself.

And please remember that I'm talking about space heating here. Commercial steam systems are different. If you're looking to cook soup in a kettle, or to make beer, or sterilize hospital instruments in an autoclave, or cook seafood, you're going to need pressure because in this case, you're looking for *temperature*. Turning up the pressure in a commercial steam system is like turning up the heat on a stove. If you have a kettle filled with water, and you need to boil that water quickly, you're much better off using steam at a higher pressure because it's hotter. Steam at 1-psi pressure, for instance, is 216° degrees. Crank the pressure up to 10-psi and the temperature goes up to 240° degrees. The water boils faster at the higher pressure, but it costs more to maintain steam at that pressure. If you own a restaurant you probably don't care. You'll just add the price of the fuel to the price of the soup.

But you don't have to heat the people in the restaurant as quickly as you have to heat their soup, and if a radiator's surface temperature is 215° that's more than enough to heat the people. The surface temperature of our bodies is only about 85°. That radiator is plenty hot for us at low pressure. If you don't believe me, go ahead and touch it.

Crank it down.

Vaporstats are lovely!

For space heating, that is. A vaporstat is like a pressuretrol, but much more accurate. It allows you to set the system to operate with ounces of pressure rather than pounds of pressure. Crank all you want on a vaporstat, the pressure won't go above 1-psi. It's like having insurance against the knuckleheads who show up on the job with those clockwise screwdrivers. They may crank, but the pressure will stay low, and the lower the pressure stays, the more fuel you'll save.

For a heating system, set the vaporstat to cut in at four ounces of pressure, with a six-ounce differential. This will operate the system between four and 10 ounces of pressure, which should be fine. If you're able to operate at a pressure lower than that, better still, and it costs you nothing to give it a try. Crank it down!

A throwback's story

A throwback is an old-school guy who took the time to learn how it was once done, and to master those ways. Steve Pajek is a throwback. Steve works with Gerry Gill in and around Cleveland, Ohio, and they both do amazing things with old steam systems. Later, when we talk about Greening the Venting, I'll tell you more about these guys, but for now, let's focus on Steve and what he did with this one job in particular.

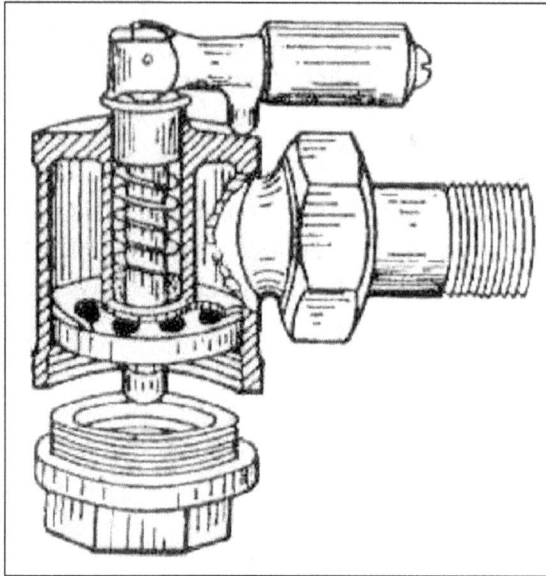

Broomell supply valve

But first, I have to tell you about the Broomell Vapor System, which the Vapor Heating Company once sold. The company was based in Philadelphia, PA, and they sold these systems all across the country during the 1920s. "Vapor" referred to the steam that wafts off an open pot of water when the water is boiling. In other words, it's steam under no pressure, or extremely low pressure. And that's what the Broomell system used. These systems had a relief valve that would pop if the pressure ever went above 10 ounces, which it rarely did because you don't need much pressure if the air is venting and the steam is of a good quality.

Broomell systems had very few moving parts. Each radiator had a supply valve with five orifices of different sizes. An orifice will allow only so much steam to pass by at a certain pressure, so the homeowner could set the amount of heat for each room by hand, which was a wonderful feature during the early days of central heating.

The return side of each radiator on the Broomell system had a small P-trap that filled with condensate and was there to stop any vapors that didn't condense in the radiator. Air passed through this P-trap and into the return by way of a small hole drilled above the water in the P-trap. The air and any vapors traveled through the dry return lines and entered a condensing radiator, which hung from the basement ceiling. This radiator's job was to kill any remaining vapors because once air passed through the radiator, it entered the chimney. Broomell used the chimney draft to induce a slight vacuum on the return lines, which helped pull the steam vapors through the supply lines to the radiators. It was wonderfully simple, and with few moving parts, so it lasted a very long time.

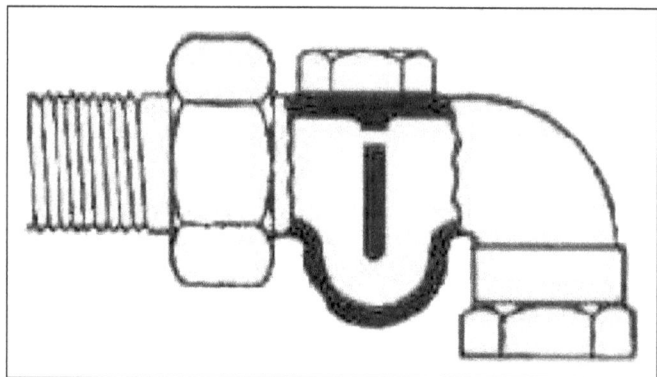

Broomell return fitting

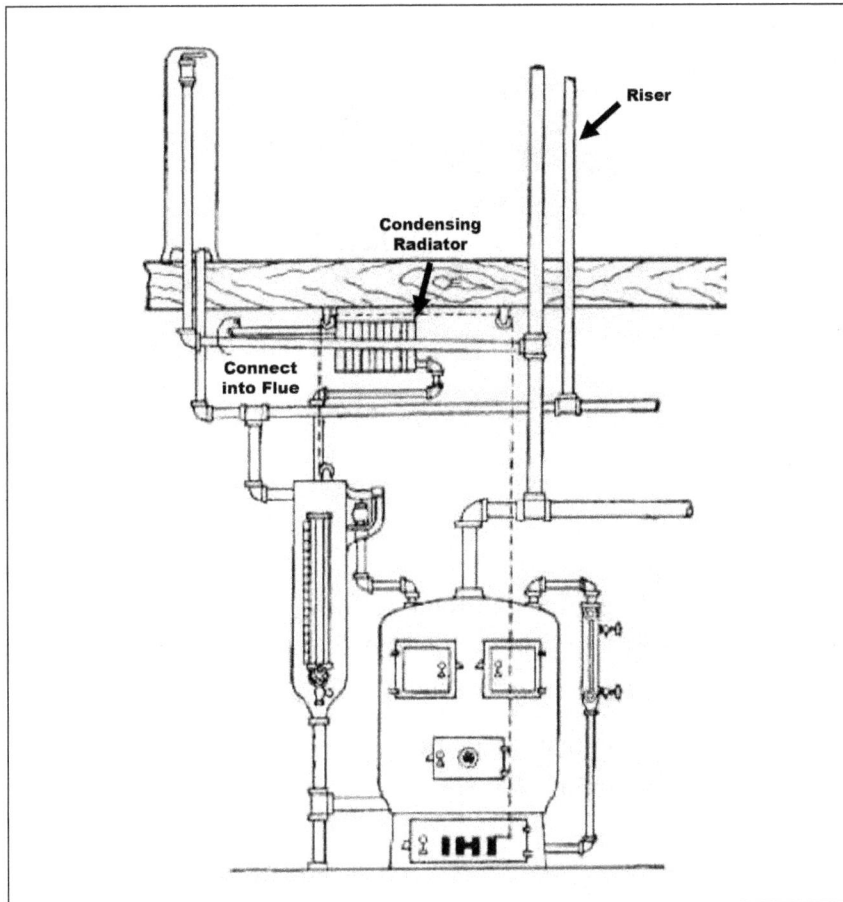

Broomell open receiver

Now here's the best part; the condensate from the radiators flowed through the dry returns, along with the air, and it fell by gravity into a large, open receiver that hung next to the boiler and above the boiler's waterline. On the side of this receiver was a gauge glass, marked in ounces of pressure, from zero to 10 ounces. Each gradation on this gauge glass is exactly 1-3/4 inches above the next. This is because one ounce of pressure represents a column of water 1-3/4 inches high. That's an important detail, as you'll soon see.

Inside the open receiver there was a copper ball, about the size of a bowling ball. It floated on the water that equalized with the waterline inside the boiler. Connected to this copper ball was a chain that ran through some pulleys on the basement ceiling. The chain connected to a draft regulator, which controlled the amount of air that could enter under the fire. As pressure in the boiler built, water would back out of the boiler and enter the open receiver, lifting the copper ball. As the ball rose, it moved the chain, which closed the damper and limited air to the fire. The fire died down and so did the pressure. The water moved from the open receiver back into the boiler and the copper ball fell, pulling the chain as it did, and allowing air access to the fire. The pressure rose again. And so on and on. With so few moving parts, this system could, and did, last for many decades, and it was able to heat very large buildings with extremely low pressure.

Nate and Steve Pajek

What made it work was very careful regulation of the flame, and very good venting. With Broomell, the chimney was the vent, and a chimney has a great capacity to move air.

Which brings us back to Steve Pajek, Throwback. Here he is with his son, Nate.

Steve is an expert at Broomell and he wanted to duplicate what that wonderful system used to do with coal, but with natural gas instead. He knew that he could save money on fuel if he could get the gas valve to move from high-fire to low-fire once the pick-up load of the system was satisfied, but the problem was there is no modern control that is sensitive enough to operate a gas valve within ounces of pressure.

So Steve built a water column that would mimic the open receiver on a Broomell system. Here's a close-up of the bottom of his receiver.

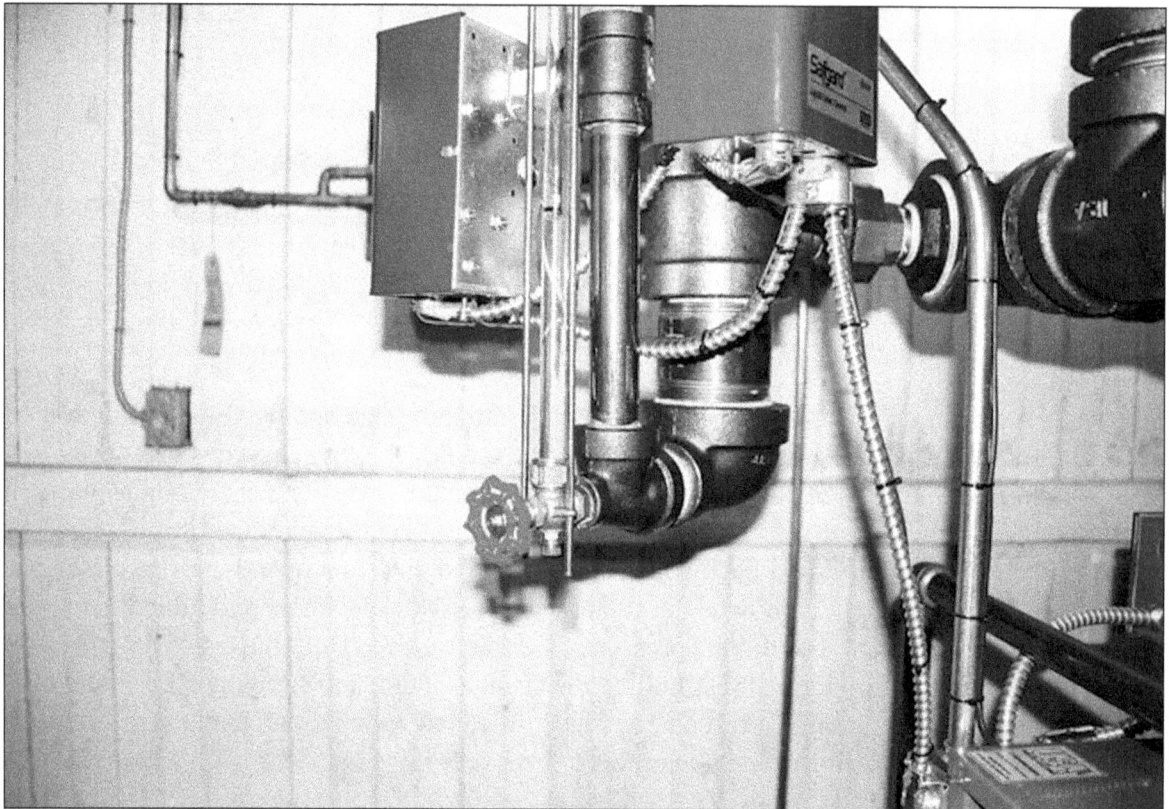

Steve's receiver

Steve is using two pump controllers from Hydrolevel. Each controller has two probes. Normally, one probe starts a boiler-feed pump, and the second probe stops the pump when the water level reaches the right point inside the boiler. Here's what it looks like:

Hydrolevel control

But Steve is not controlling a pump. He's using the two controllers to start and stop the burner, and to cycle it between high- and low-fire while it's operating. With two controllers, he has a total of four probes to do this. So follow along with this photo as I describe the operation:

One of the pump controllers opens a circuit when the water reaches its top probe. The other closes a circuit when the water reaches its top probe.

Steve set it up so that the highest probe opens the circuit. This is the operating limit (the cut-off point). The next probe down is the top probe of the other pump controller. It will close when water reaches it, and it is the trigger for the low-fire gas valve.

The next one down is the lower probe of the latter-mentioned unit. This is the ''release'' probe to disengage low-fire and reactivate high-fire.

Continuing down, we have the next probe, which is the lower probe of the first-mentioned pump controller. It is the release probe for that unit, so that when the limit probe activated, this probe releases that activation bringing the boiler back on.

The burner starts on high-fire, and as the boiler pressure rises, water will leave the boiler and enter the reservoir that Steve built. It will rise above the lower probe, and then the second probe, and when the pressure is high enough to have the rising water reach the third-highest probe, it will energize the low-fire valve.

Naturally, with less heat going into the water, the pressure will begin to drop and the water will begin to equalize back into the boiler. When the water in the reservoir reaches the second probe, it disengages the low-fire valve and restores high-fire. If the pressure didn't drop when low-fire was engaged, but continued to climb anyhow, it would hit the fourth probe, which shuts off the burner. The water then falls until it clears the first probe, which restores the burner to operation.

You space your tees based on 1-3/4" of column height equaling one-ounce of steam pressure. Steve tells me that it's *very* accurate.

And this is how you get a boiler to simmer away between high- and low-fire at extremely low pressure, just as it did in the days of coal. I think it's brilliant. And it sure is green.

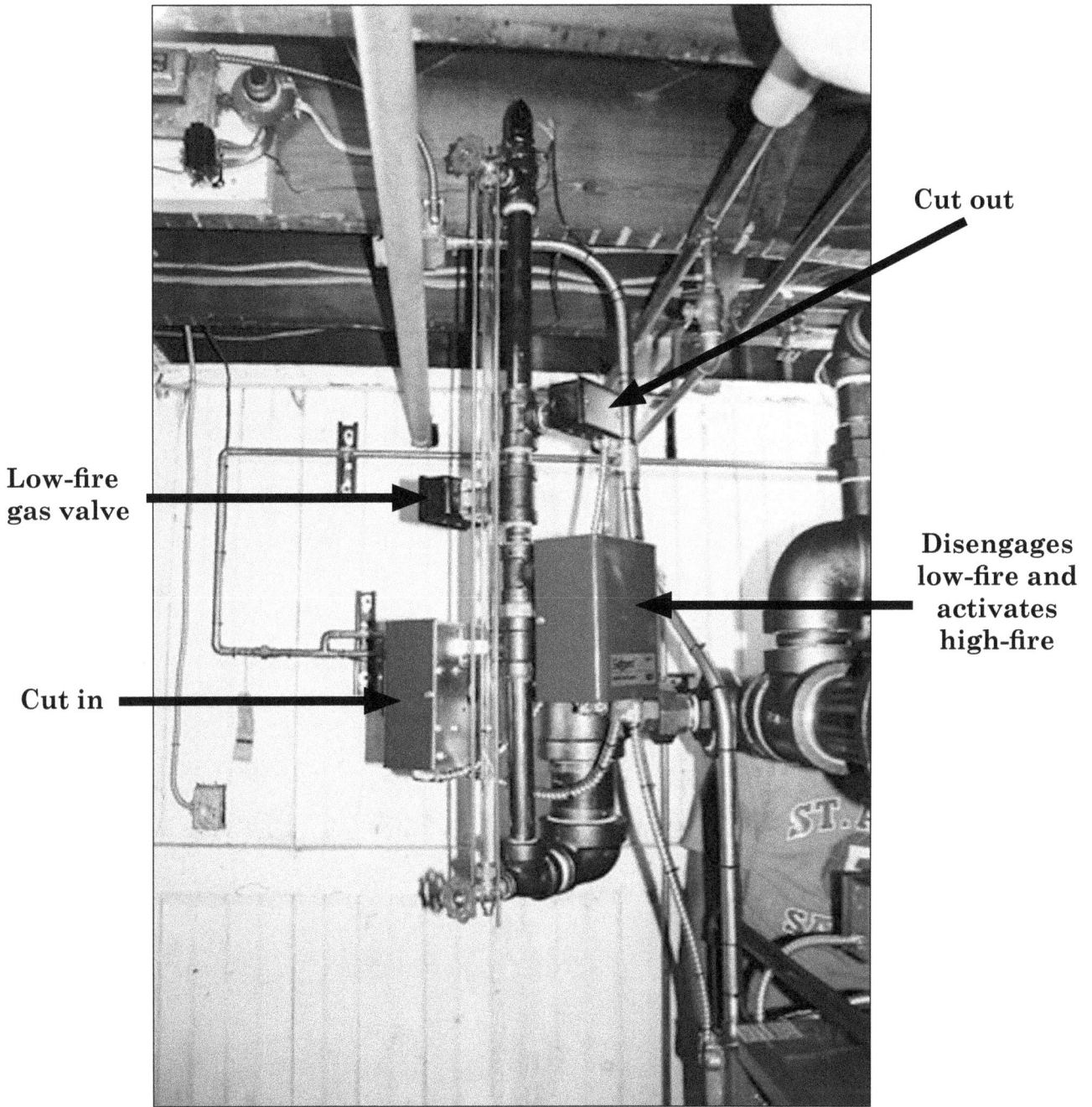

Cut out

Low-fire
gas valve

Disengages
low-fire and
activates
high-fire

Cut in

Steve's receiver

GREENING THE WATER

If the water isn't right, the steam won't be dry.

Dry steam is a technical term. It means that the steam leaving the boiler contains no more than two percent liquid water. That's what's best for the steam heating. More liquid than that will probably cause much of the steam to condense before it reaches the radiators. The steam gives up its latent heat energy to the liquid that it's carrying along with it, and winds up carrying fewer Btus to the radiators. The result? You paid good money to make the steam, but it died before it had a chance to warm the people. That's not good for the budget, so let's spend some time talking about fixing the water.

Here are the main things that go wrong with boiler water:

- It can contain oil, so we'll look at where that oil comes from.

- It can be too hard (which is also hard on the budget).

- It can contain too much acid or too much base, and when it comes to pH, we're looking for the Goldilocks approach – not too much, not too little, but *juuuuust* right.

- It contains too much dissolved solids, which is a fancy way of saying it's just plain dirty.

Okay, let's take them one at a time.

Oil

Oil floats on water because it's lighter than water, but where does it come from? How does it get into a boiler and what does it do to the steam quality?

Good questions. Let's begin at the factory where the manufacturer made that boiler. If it's a cast-iron, sectional boiler, the manufacturer had to have access to a foundry. They heated iron until it liquefied and then poured it into a mold. The iron flowed around a sand core and hardened as it cooled. The manufacturer shook out the sand and now had an unfinished, hollow boiler section. Each section is like a slice in a loaf of bread. The manufacturer holds the sections together with tie rods, but first, those sections need some machining so that they fit together properly, and they also need threaded holes so an installer can add pipes to the boiler.

Visit a boiler manufacturer and you'll see these huge machines that drill and tap the holes in the iron castings. To keep the tools from bursting into flames, the machine lubricates the casting with oil as the drill operates, and that oil flows freely. The manufacturer doesn't clean the casting afterwards. The oil coats the iron and keeps it from rusting as much as it naturally would without the oil. The casting looks better with the oil on it, and since iron is porous, the oil finds lots of nooks and crannies in which to hide. It's in there, waiting for you.

The installer also adds oil to the water because, to thread pipe, you need cutting oil. Few installers take the time to clean their pipes with soapy water before hanging them on a boiler. That oil winds up in the water as soon as the boiler starts making steam, and this affects the quality of the steam. Think about all the new pipes that go around a replacement steam boiler. There's a lot of cutting oil involved in that work, and it's going to wind up in the water. This is one of the reasons why a replacement steam boiler often has problems that the old boiler didn't have.

A big building may have a boiler that burns Number 6 oil. We call this "heavy oil" and it costs less than Number 2 light oil, but it's also impossible to move Number 6 oil at low temperature because it's like molasses. You have to preheat heavy oil to loosen it up before you can burn it, and the fuel-oil preheater is often a steam coil that's inside the big oil-storage tank. That coil can become a source of oil in the water if all the coil's piping connections aren't tight, or if the coil should spring a leak.

Once oil is in the water, it will emulsify with alkaline boiler water, which causes the water to prime and surge. This creates wet steam as the liquid water begins to leave the boiler with the steam. The water robs the steam of its latent heat, causing the steam to condense before it reaches the radiators, and that wastes fuel.

Try an experiment with two pots of tap water. Add some cooking oil to one of the pots, and leave the other with just the plain tap water. Put both pots on the stove and turn the heat on high. Watch how each boils. The pot with the mixture of oil and water is going to boil with much more violence. The water will be rocking up and down and probably spilling out of the pot. The oil is causing that, and chances are the water isn't even that alkaline (that means the pH is higher than 7).

And by the way, this is also a great way to convince a customer that the water in the boiler needs attention. If you suspect that there's oil in their boiler, take a sample in one of those pots. Put tap water in the other pot. Boil both samples and watch what happens if the boiler water contains oil. Seeing is believing; and your job is to get rid of that oil.

Getting rid of the oil

The best way to get rid of the oil is to skim it from the surface of the waterline. Does the boiler you're working with have a skim tapping? Look for it near, or just above the waterline. It should be a relatively large tapping in the side of the boiler if you're working with a commercial boiler. If it's a small, residential steam boiler, it may not have a skim tapping, which is going to make things more difficult, so if you're in the market for a steam boiler, look for one that has this tapping.

Don't waste your time skimming from the tapping in the top of the boiler. That's the main steam connection that goes to the header (more on this when we get to the chapter on Greening the Piping). Imagine yourself as the oil floating on the water's surface. Someone is trying to skim you from a top tapping on the boiler. As the water rises, you, as the oil, just cling to the top of each cast-iron boiler section and refuse to leave. You have so many places to hide up there.

It's like a diver trying to come up through a hole in the ice, without knowing where that hole is.

This is why side-skimming works and top-skimming doesn't. You have to float the oil out horizontally, just as you would skim grease off the top of spaghetti sauce.

Here's how to do it.

Pipe a full-size nipple and a full-size tee into the horizontal skim tapping. You're using a tee so you'll be able to watch the water as you skim. The side tapping of the tee acts like a view port. You'll see where the waterline is. Without the tee, you're liable to be skimming from below the surface of the water, and the oil will just remain in the boiler. After the tee, use an elbow to point the water in the direction of a bucket. When you're done skimming, add a plug to the bull of

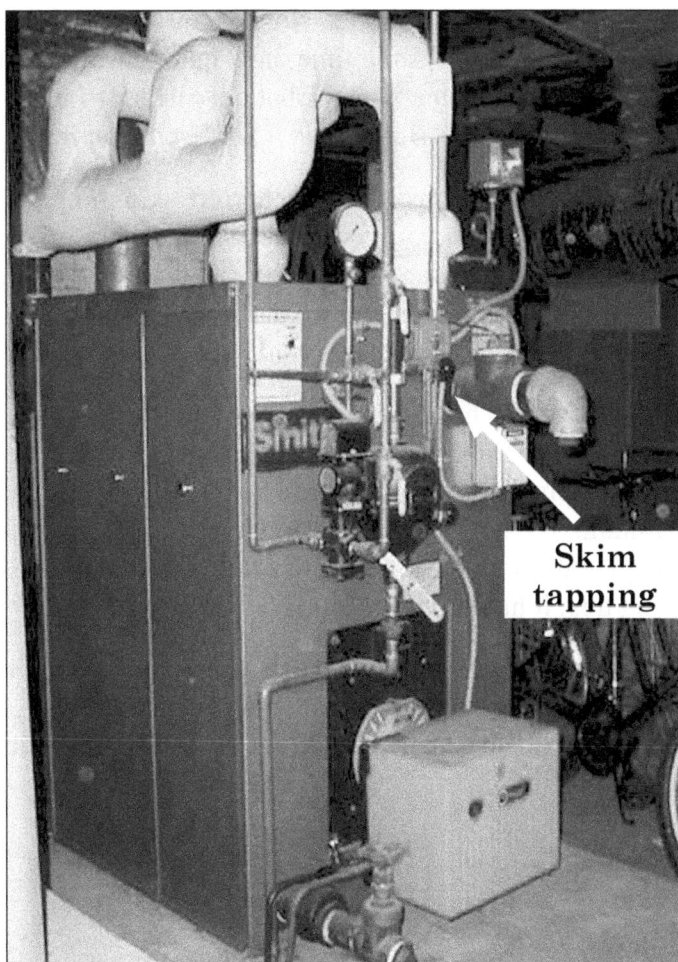

Skim tapping

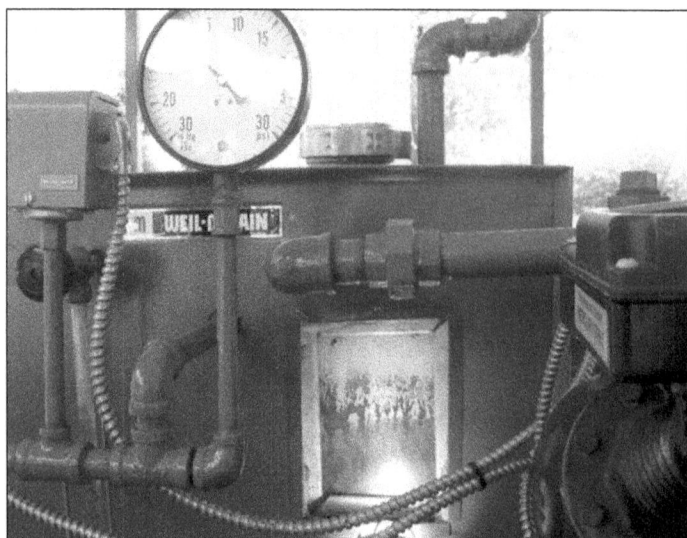
Weil-McLain's porthole and gauge glass

the tee and to the elbow, as you see in the photo. You're done.

After the installation of a new or replacement boiler, let the burner run for a few days before you skim. This gives the oil that may be on the pipes a chance to work its way back into the boiler. If you wait a few days, you'll catch more of it.

In their instructions, some boiler manufacturers say to heat the water while skimming, but not to let it boil. I find this to be very difficult because the waterline moves a lot when the burner's firing. Here's an example of that in the photo above.

The good folks at Weil-McLain have a special boiler that they bring to tradeshows. It has glass piping so you can watch the steam and the water, and it has this porthole so you can see the waterline inside the boiler. Notice how violent that water is (I wish you could see it in real life), but note, at the same time, how quiet the water in the gauge glass is. When I took that photo, it was bouncing slightly, but not much. If I were on a job with that boiler I would have thought all was well. But look at what's going on inside that boiler.

This is why I think you're better off skimming when the water is hot, but not when the burner is on. When you get to the job, just shut off the burner, open the plugs on the skim fittings, feed the boiler slowly with street water and watch the waterline through the bull of the tee. Don't be in such a hurry that you're skimming from where the oil ain't. And please be patient. This is going to take a while.

How long? That depends on how much oil is in the water. A way to check is to use a test pipe. This is a trick an old-time oilman on the Isle of Long taught me years ago. Get yourself a 2" X 10" nipple with a cap. Clean both really well so that there's no oil on them. While you're skimming the boiler, catch a sample of the water in the test pipe, and then use your torch to bring the water in the test pipe to a boil. Oh, and please don't hold the pipe in your hand while doing this. It hurts.

Look into the test pipe with a flashlight as the water boils. Is the water priming and surging? If so, it probably still contains oil, so keep skimming. Be patient; it often takes a full day to properly skim and clean a steam boiler, but it's worth it because dry steam saves fuel dollars.

Chapter 4: Greening the Water

Skim tapping

The photo above is another example of a skim tapping.

This boiler has side outlets for the steam, one of which can double as a skim tapping. Follow the same procedure I just explained.

And notice the small fittings and ball valve at the top of the boiler (to the right in the photo) that this installer used to make a funnel. He's now able to add some strong soap to the boiler when he cleans the system. I'll tell you more about that in just a few minutes.

And then there's mud

Or you might call it sludge; it's your choice.

At the bottom of every steam boiler there is a mud leg, which gives the mud that forms a place to settle. Water circulates through the mud leg on its way to becoming steam, and if mud slows or stops that circulation, parts of the boiler might overheat and fail.

To keep the boiler from suffering an early death, you need to flush that mud leg. How often you need to do this flush depends on how much mud collects, so check to see how much feed water is entering the boiler. That's mainly what brings in the mud.

This mud isn't like what happens after a heavy rain; it's a goop formed by a combination of scale-forming minerals that may be (and probably are) in the feed water, iron-oxide corrosion products (which is a fancy way of saying rust), and treatment chemicals.

If the boiler serves a large building, it's good to work with a water-treatment company. They'll probably treat any mud-prone boiler with dispersants, which prevent the mud from sticking to the boiler's heat-transfer surfaces, making it easier to get rid of with a good flush.

To blow down the mud and get rid of it, you'll need to have certain valves around the boiler. We'll talk more about this when we get to the Greening the Piping chapter. For now, let's stick with the water and all the wacky things it can do when it's not just right. For instance:

Priming

This is the carrying-over of water droplets with the steam. Think of priming a pump. Priming is bad because it robs the steam of its latent heat energy, causing it to stop moving out to the radiators. Priming costs money.

Surging

Surging is the bouncing of the water level inside the boiler. You can see it in the gauge glass. And when it's happening in the gauge glass, it's also happening inside the boiler – and usually much worse. Surging can cause the low-water cutoff to quickly start and stop the burner and that wastes fuel because the burner never gets to its steady-state stage of firing.

Foaming

You can sometimes see foaming in the gauge glass. It's those bubbles that form in the glass above the waterline. It's tough for the steam to work its way though this surface foam. The water often goes out into the system with the steam, and that wastes fuel as the steam gives up its latent heat to the water.

When the boiler water's pH is too high, the water will foam, so let's take a closer look at pH.

The power of hydrogen

You probably learned about this in grade school. Your science teacher introduced you to litmus paper and let you test lemon juice and baking soda and whatnot. The juice was acidic and the baking soda was alkaline. One showed a low pH and the other showed a higher pH. The scale went from zero to 14 and the number seven was neutral. Here's that scale.

When we measure pH we're measuring the **p**ower of **H**ydrogen. The number refers to the relative concentration of hydrogen (acid) and hydroxide (base) ions in solution

For a steam boiler, a pH between 7 and 9 is just right. If it's lower than that, the water will start eating the pipes. If it's higher than that, the water won't corrode metal, but it may begin to foam.

Hold that thought for a moment.

Let's say you have a boiler that's taking on a lot of feed water because the system has leaks. You should fix the leaks, of course, but it seems to be cheaper to keep adding feed water (it's not), so that's what's been going on.

Increasing Acidity	0	
	1	— Battery Acid
	2	— Lemon Juice
	3	— Vinegar
	4	
	5	
	6	
Neutral	7	— Milk
	8	
	9	— Baking Soda, Sea Water
	10	
	11	— Milk of Magnesia
Increasing Alkalinity	12	Ammonia
	13	Lye
	14	

The feed water brings with it carbonates and bicarbonates, which are both natural and normal. The carbonates and bicarbonates form carbon dioxide when the water boils, and this gas flows out into the system with the steam. When the steam condenses, it absorbs the carbon dioxide and turns into carbonic acid. This is bad news because the carbonic acid will remove the thin film of rust that naturally forms on the insides of steel pipes. That mild rusting is a good thing to have because it helps protect the underlying metal from further corrosion. By stripping away the surface rust, the acid makes more fresh metal available for munching.

This is why it's good to insulate the return lines in any steam system. The cooler the water gets, the more it will absorb gases such as carbon dioxide. Keep the condensate as hot as possible and it will be less acidic.

To avoid a low pH and the corrosion that comes with it, many people add chemicals to their boilers. These push the pH up toward the alkaline side of the scale. When the pH reaches 10, corrosion becomes impossible. Get a bottle of drain cleaner and check the pH. Note how high it is. That's why they can make that "It's safe for pipes!" claim on the label.

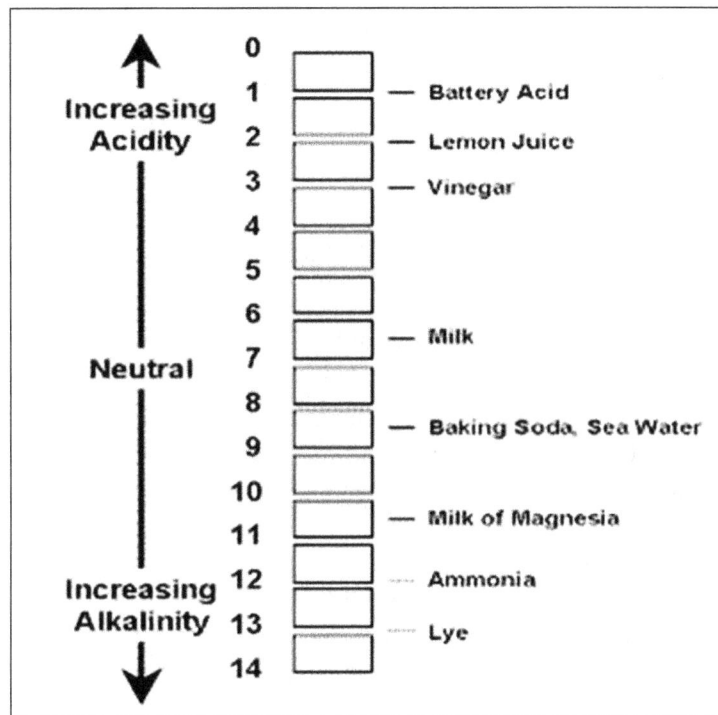

But if the pH reaches 11 (and this often happens when you add too much anticorrosion chemical to the water), the water will foam, and that leads to wet steam. Wet steam wastes fuel because it robs the steam of its latent heat. It stops before it can reach the radiators. You wind up running the burner longer, and you still get lousy results.

The classic off-the-shelf chemicals used to adjust pH are baking soda and vinegar. Check out where these are on the pH scale and you can see why the Dead Men used them to raise and lower the pH. They're inexpensive and they work, but if you use vinegar (or any other chemical), beware of the odor it will send upstairs in a building heated with one-pipe steam. The air leaving the radiator vents will be smelly, and that can lead to complaints.

It pays to check the pH when you're blowing down the low-water cutoff. Keep some litmus paper in your toolbox or near the boiler. It's a simple test and easy to do, and by getting the water's pH just right, you can solve a lot of problems that are costing money.

How much feed water?

What's a normal amount of feed water for a steam boiler? Good question. All steam systems are open to the atmosphere, so there's always going to be some evaporation. But steam boilers that do only space heating reuse nearly all of the water because the condensate returns to the boiler from the piping and radiators. You shouldn't have to add too much water to these systems unless they're leaking. And it's always better to fix the leaks rather than to keep adding fresh water.

The trouble with feed water is that it's cold and it contains lots of oxygen. Industrial boilers deal with that oxygen by passing the feed water through a deaerator, but you'll rarely see a deaerator on a space-heating boiler.

Henry's Law tells us that gases dissolve in liquids in direct proportion to pressure and temperature. So the colder the water, the more oxygen it will contain. That oxygen comes out of solution as the water boils, and it can eat holes in the boiler, right at the boiler waterline.

Feed water also contains suspended solids. The more feed water you allow in, the more solids you're going to get. The solids collect on the surface of the water as it boils. They surround the steam bubbles as they form, making them tougher. Tough bubbles resist breaking, and that leads to foaming. The finer the suspended particles are, the more they will collect in the bubbles, and the worse the foaming will be. You probably can't see these solids but they're there, and this is why you must keep the feed water to a minimum. Fix those leaks.

So once again, what's normal for feed water? Well, I think it depends on that system's history. Has anyone kept track of how much water has entered over the years? Is there a meter and a logbook? Are the pipes leaking now? Are they buried under the floor? Do they go through walls? Are they in places you can't see? Are the main air vents in crawl spaces where they may be leaking steam, but go unnoticed?

"Normal" is what traditionally happens from day to day. Have there been sudden changes? How would you know unless you've kept records?

I think every steam boiler that has an automatic water feeder should also have a water meter. Some of today's electronic automatic water feeders incorporate a water meter into their circuitry, and these are worth the money because they tell you when something changes. If the system suddenly starts taking on lots of water, you know there's a leak, and it's either in the pipes, or it's inside the boiler. Read on.

Holes in the boiler

In 1973, the first oil embargo hit America. The price of fuel soared and many people looked to get rid of their old coal-converted steam boilers. They wanted something more efficient and the boiler manufacturers met the public's need by introducing smaller steam boilers with higher efficiencies. They fired these little boilers hard, and the steam left them much faster than it used to leave the old boilers. And the steam left through much smaller holes, holes that the Dead Men never would have considered wise. I know this because they wrote about it in their books. In the old days, there was never a steam boiler with an exit hole smaller than three inches.

Many of these boilers that arrived in the '70s were actually hot water boilers, trimmed for steam. I don't think the people making the decisions at the time fully understood what made for a good steam boiler. It's understandable, though. Steam heating had been a profitable, non-growth business for decades, and many of the engineers who worked for the boiler companies during the '70s had never designed a steam boiler. They had been selling the designs of the Dead Men, and everything had been fine up to that point, but now they had to come up with something small and efficient, and in a hurry. And that's when the troubles began.

I remember one oil-fired steam boiler from the late-1970s that had very narrow sections. The water inside those tight sections would fill with steam and rise to a level higher than the level in the gauge glass. When the boiler reached high-limit on the presssuretrol's setting, the burner would stop, which caused the steam that was rising through the boiler water to collapse. This drove the waterline inside to boiler to a point lower than the waterline inside the gauge glass. And as the different water levels tried to equalize, the automatic water feeder would kick in. Within a day or so, the boiler would be flooded.

From the outside, it looked like the water was backing out of the boiler, so the installers started adding check valves to their wet returns, but this didn't help. I visited many of these jobs. The boiler was creating the problem, but the contractor and the homeowner were holding the bag.

The manufacturer stopped making this boiler, and that was a good thing. But before that happened, many of the installers ordered larger-than-needed boilers and then down-fired them. This lessened the amount of steam in those narrow boiler sections, and it seemed to solve the problem, but it often left the homeowner with a less-efficient boiler.

But I digress. What does this have to do with holes in the boiler?

Years went by, and better steam boilers arrived – ones with wider sections, larger (and more) openings that allowed the steam to leave at a lower velocity, and bigger steam chests. But even some of these boilers began to fail within a few years, and many from holes at the waterline.

I'd hear about the holes from contractors. They'd tell me the boiler was only a few years old, but now it had a hole in it, and right at the waterline. Some talked about more than one boiler suffering the same fate.

I went back to the old books and confirmed that the leading cause of this is the oxygen brought in with too much fresh feed water. Every steam system is an open system, and excessive oxygen will always cause iron to rust and eventually fail. If the piping system is leaking (and many old steam systems have those buried returns), the automatic water feeder will do its job regularly, and that can lead to a hole at the waterline. But many of these jobs had water meters on the feed line. There were no significant leaks in these systems, which made me wonder.

Then I met a rep from a boiler company at an industry event. He told me that they had figured out what was causing the holes in the boilers on those systems that weren't taking on a lot of fresh water. This company had hired an independent lab to analyze the metal of boilers that had developed holes at the waterline. The problems were clustered in a geographical area, which made the company and the laboratory suspect that there was something in the water in that area that was causing the problem. It turned out to be high levels of chloride.

Now this is not chlorine, which some water companies use for treatment; this is *chloride*, or to put it more simply, salt. Chloride can come from the runoff after deicing trucks treat the roads during the winter, or it can come from natural sources. But once it gets into a boiler, it can cause problems, particularly if it comes in contact with abnormally hot surfaces.

Most of the boilers we use for steam heating nowadays have cast-iron sections with pins for efficient heat transfer from the flame to the water. These pins are also in the upper part of the boiler, up there where the steam disengages from the water. We fire these boilers hard to get the most steam from a relatively small appliance. And that's where the problems start.

The combination of the chlorides in the water, and the very hot pins in the steam section of these modern boilers (it's too bad they have to be so hot), creates a condition known as graphitic corrosion. You can Google that and learn more, but if you're not up for a science lesson, just know that graphitic corrosion eats cast iron. You wind up with holes at the waterline through which you could toss a tomcat.

The high temperature is the trigger that starts the corrosion process and those pins are as hot as they are because modern, vertical-flue, pinned-section steam boilers are more efficient than the ones they replace. But if you look at the boilers from the old days you'll notice that none of them had pins. They had smooth, ribbed surfaces and a lot more metal, and because of this, the older boilers weren't as sensitive to chlorides in the water as modern boilers are. The ones we take out were less prone to developing holes than the ones we put in.

Some manufacturers tried to solve this corrosion problem by using more boiler sections for the same Btuh rating. This lowered the temperature of the boiler's above-waterline, cast-iron pins, but it also made the manufacturers who tried this tactic less competitive. And, as it turned out, it only postponed the graphitic-corrosion problem. The holes still formed.

The chlorides are the problem, and the excessively high temperature on the pins is the trigger. The problems are widespread, occurring in the western and southern suburbs of Boston, Massachusetts, the northern and the northwestern suburbs of Providence, Rhode Island. The problem is also in New York State, on the eastern side of the Hudson River, from Albany to Peekskill, and in many parts of Long Island, NY. These are all places with lots of steam heat.

So if you're using steam boilers of this type and you're having problems with corrosion, look to the water. Are you taking on a lot of feed water? The only way to know for sure is to meter it. Is the pH of the water between 7 and 9? The only way to tell is to test it and to monitor it.

Once you've checked those things, have the water tested for excessive amounts of chlorides. That could be the reason why the boilers you're taking care of aren't lasting. And this sort of internal corrosion isn't covered by most boiler manufacturers' warranties, and neither is the labor involved in replacing these boilers once they fail.

Does your system need water treatment?

If the water contains chlorides it sure does, but steam heating is different from steam process. Nearly all the water in a space-heating steam system gets returned to the boiler and recycled into steam. Less feed water means less mineral deposits will enter the system, so less scale forms on the hot boiler surfaces

Steam-heating systems do take in a lot of oxygen, though. They're open to the atmosphere through main- and radiator vents, and the vent lines on condensate- and boiler-feed pump receivers. Every time a steam-heating system starts and stops, it breathes air in and out. There's more of a chance for oxygen to cause corrosion in a steam heating system than there is in a commercial steam system because commercial systems stay on most of the time. They don't go through the on-off cycling that a steam-heating system goes through.

You should consult a chemical-treatment company if you're concerned about this. They'll probably suggest adding an oxygen scavenger to deal with the feed water. The ones most commonly used are sodium sulfite and hydrazine. Both do a great job of absorbing oxygen.

They'll probably also suggest a corrosion inhibitor. The one used most often in steam-heating boilers is cyclohexylamine, which is a neutralizing amine. It neutralizes dissolved carbon dioxide so that the protective coating of iron oxide can form inside the boiler, and inside the supply- and return pipes.

Try to find out what's in the commercial chemical products you use. Don't just buy it because it has a cool name.

Mixing MEX

There's a company on Long Island called Meenan Oil. They got their start in the late '40s with Levittown, the world's first tract-housing development to use radiant heating. Meenan is now part of a larger company, but when it was an independent, a fellow named Howie Straub worked there, and he was one of my teachers. Howie was old enough to be my father and he was always willing to share what he knew. I used to stop by with my questions and Howie always found time for me. I miss him.

One day, I was asking him about dirty steam boilers and he showed me a box of MEX, which is a commercial soap that you'll find in most hardware stores. Howie explained how they used to use trisodium phosphate (TSP) back in the day, but the phosphates hurt the aquifer, and now the local government had banned them. Howie said that MEX works just as well as TSP.

He showed me how to mix the MEX with hot water in a bucket. We wouldn't need much – just one pound for each 50 gallons of boiler water, and since modern boilers don't contain much water, a few scoops should do it. He also told me that I could get the water content of any boiler from the boiler manufacturer.

We'd pour the mixture into the boiler through a funnel that we'd place in an upper tapping. Sometimes we'd have to remove the relief valve to do this. Other times, we'd build a funnel from fittings, just like the one in that photo I showed you earlier when we were talking about skim tappings.

We'd let the MEX run through the system. It would show up back on the return side, and we'd trap it with a shut- off valve and let it run out the drain valve. I'll show you more about this piping in the next chapter.

The MEX did a great job of cleaning not only the boiler, but also the system's pipes. In fact, some of Howie's customers would call to complain that their domestic hot water was now too hot because the MEX had cleaned the outside of the tankless coil in their boiler. Howie would have to go back to adjust the water temperature.

When the returns ran clean, we'd make sure that we flushed all the MEX from the system so that the boiler water wouldn't foam afterwards. It took a while but it worked every time, and whenever I tell about it, I think of my friend, Howie Straub. He understood the real world, and he was a fine teacher.

I sure do miss him.

Time is money

If you're a contractor, you're going to have to figure the time you need to properly skim the boiler. You'll get a better sense of the time involved by doing it a few times, but from what I've seen, it takes at *least* a full day to do it properly.

Skimming oil and cleaning the water will save fuel because the steam will be drier. If you explain this to your customers whenever you're replacing a boiler, they'll realize that having you clean the boiler is in their best interest. You'll tell them what happens to a steam system when the water isn't right. How the burner will run longer. How water will carry over into the system, robbing the steam of much of its latent heat, and how it will often cause water hammer in the pipes. You'll tell your customer about the damage water hammer can do to a steam system. If they're reasonable, they'll listen.

If you're replacing the boiler, I suggest you include the cleaning as an addendum to your boiler quote. That way, if your competitor doesn't include it, you'll have a good talking point. If you include the price of the time involved for a good cleaning in your boiler quote your price may appear to be too high compared to your competitors. So break it out and tell your story.

If your customers are smart, they'll go for the addendum and you won't be doing this work for free. If they don't want the cleaning, have them sign off on it. Make up a form that explains what happens to a steam system when the water is dirty and full of oil. Show them these pages. That way, when the new boiler acts up, you'll have something to fall back on – you told them the system and the boiler needed cleaning. They refused to let you do the right thing.

And keep in mind you may have to clean the boiler more than once. Experience will teach you better than I can. Talk about this stuff up front and you'll get along much better with that customer. And you'll get to keep more green in your wallet.

A note on commercial applications

This has nothing to do with steam heating, but it's something you should know if you're a heating contractor. If you're using a steam boiler in a bakery to make rolls crusty, or to cook seafood in a restaurant, you're probably wasting all of the condensate and bringing in 100% feed water. The oxygen in the cold feed water is going to eat that boiler for lunch, and the people who own that boiler think this is normal, which it is, but there's something you can do to increase the life of those boilers, and that's going to make you look like a genius.

In addition to the oxygen, which is eating the metal, the minerals (the hardness) in the feed water will settle onto the boiler's hot surfaces. That adds a layer of insulation inside the boiler and the metal will begin to overheat. Eventually, it will fail.

So here's what you should do: Get a small domestic water heater and use it to feed the boiler. Set it on its highest temperature. The mineral deposits in the cold water will settle out inside the water heater instead of inside the boiler. You'll replace the water heater, for sure, but small water heaters are a lot cheaper than steam boilers.

Next, get a good hydronic, microbubble separator and pipe it on the hot-water line between the water heater and the steam boiler's automatic water feeder. As the feed water flows, the air separator will toss out most of the oxygen that's in the hot water so that it can't corrode the steam boiler.

It's another one of those old-timer tricks. A water heater and an air separator make for a fine, low-cost deaerator for that light-commercial job. And as I said, *you* get to look like a genius.

Go get 'em.

CHAPTER 5

GREENING THE PIPING

What makes piping green?

I mean, it's just there, right? It may have been there for 100 years. So all of a sudden it's supposed to be green? What's that all about?

Well, if we go back to what we agreed on at the start of this book, "green" has to do with the money in your pocket, as well as with the environment. The piping in that steam system, particularly the piping near the boiler, has a lot to do with it. It may have been okay to start, but how many people have made how many changes over the years? And how might those changes affect the boiler's ability to produce dry steam?

How many boilers have served this building since its doors first opened? How many people have worked on the piping? How many of those people made things up as they went along?

You don't know, do you?

I don't know either.

Steam boilers have changed over the years, and the new ones often don't get along with the old piping. So let's take a close look at all the pipes that carry the steam and the condensate, and let's figure out how to make that old system more efficient. Let's make it green.

Out with the old, in with the new

What made a steam boiler a steam boiler? Well, first of all, it had big, wide internal sections that allowed plenty of room for the steam to form and rise through the water without causing the water to rock violently. Just think of steam moving through a wide channel, and then imagine the same amount of steam moving through a channel with a quarter (or less) of the width. It's going to get crazy in there. The tighter the space, the faster (and more violent) the steam.

A true steam boiler also had a large area to collect the steam as it rose from the surface of the water. Some old boilers had big steam drums that sat atop the boiler. These cathedral-like spaces gave the steam a place to slow down and drop its carryover water back into the boiler. This made for drier steam and lower fuel bills.

A true steam boiler had big outlets for the steam, and the manufacturers designed these so that the steam was never moving faster than a certain critical velocity (usually 15 feet per second). If the steam moves too fast as it leaves the boiler, it will pull the water out of the boiler, and that leads to wet steam. Back in the day, steam boilers never had exit holes smaller than 3" in diameter. Today's steam-heating boilers have holes as small as 2". This makes a difference in the quality of the steam.

As I mentioned earlier, the steam boilers began to change after the 1973 OPEC Oil Embargo. It was our first energy crisis and people scrambled to get rid of those old, fuel-guzzling boilers. That was our first serious Green (as in $$) Movement.

The boiler manufacturers responded to the consumers, and to the government, which was demanding higher efficiencies, by making steam boilers smaller. Residential steam boilers got *really* small.

In some cases, hot-water boilers became steam boilers. All the manufacturer had to do (or so they thought) was to change the trim on that little house-heating boiler. Remove the circulator and add a gauge glass. It seemed simple and economical because, by the mid-1970s, steam had become a profitable, non-growth business for the manufacturers. It didn't make sense to invest in R & D for new steam-boiler development. No one was installing steam heat from scratch anymore. The entire market was replacement boilers.

Those new boilers had a higher combustion efficiency than the ones they replaced, but they also had narrow internal sections, smaller steam chests, and smaller openings (some as small as 1-1/4"). Most produced very wet steam, and that resulted in higher fuel bills. It wasn't a happy time.

The years passed and the boiler manufacturers fixed most of these problems, and they did it by coming up with specific instructions for the piping around the boiler. They learned that this near-boiler piping was the key to making the steam dry. It could remove most of the carryover water from the steam before the steam headed out to the radiators. From that point, success or failure was in the hands of the installer. Would he or she follow the piping instructions? Or would the urge to take shortcuts win out?

Let's look at some practically perfect piping.

All around the boiler

This is where you'll remove the water and make the steam green. I've had this drawing for years. It's a cast-iron sectional boiler, the sort you'd have in an apartment building, and everything in that drawing is just about right.

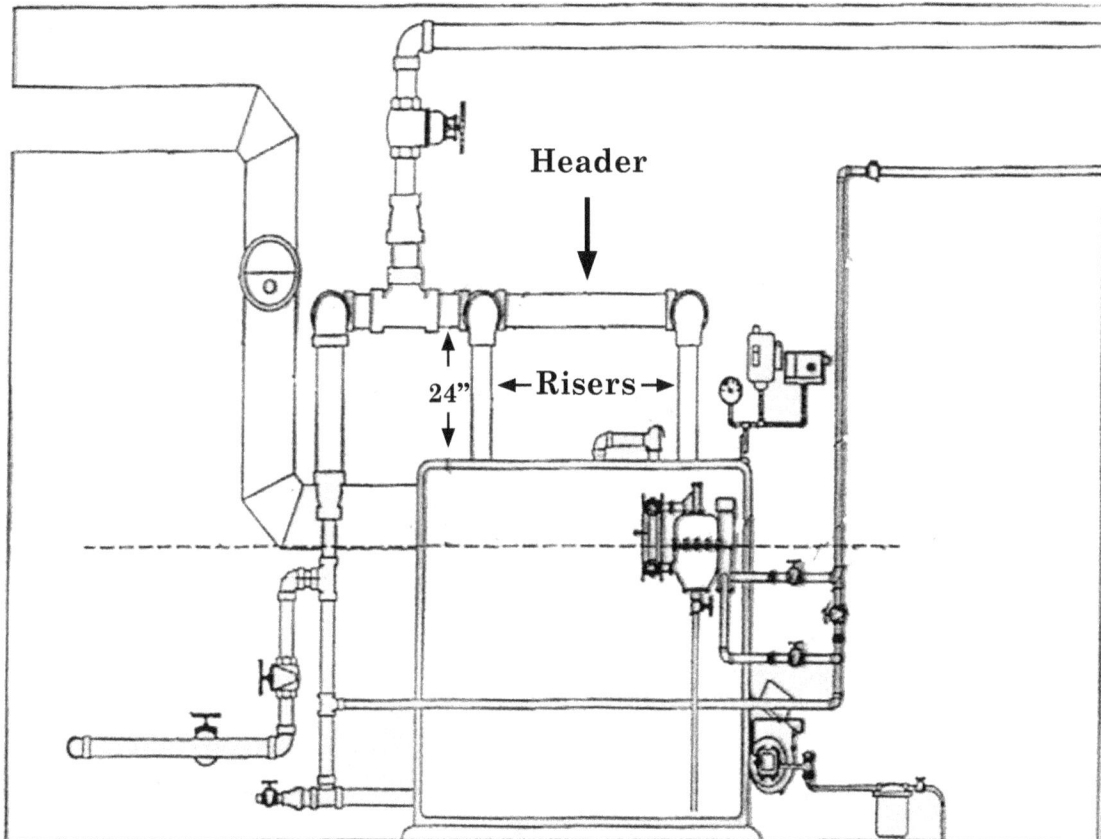

Near-boiler piping

Let's start with the exit holes on the boiler. The size of the hole is going to play a big part in how fast steam moves out of the boiler, and the slower it goes, the less likely it is to take water with it. That means, the bigger the hole the better. In the case of this boiler, we're using two holes. That cuts the exit velocity in half.

And this brings up a good point. I think you should use all the holes you can to make the steam go as slowly as possible at this critical point. If you buy a boiler and it has more than one outlet, use them all. The more holes you use, the drier the steam will be.

Now, here's something to consider, and it goes on in the field all the time. Let's say Boiler Manufacturer #1 makes a boiler of a certain size and he feels that two outlets are the way to go for this size boiler because he's looking to get you the driest steam possible. Boiler Manufacturer #1, however, has to compete with Boiler Manufacturers #2, #3, #4, and all the others.

Suppose Boiler Manufacturer #2 decides that the steam is going slow enough with just one outlet, even though this is questionable? His boiler of a certain size comes with two outlets, but he decides to tell you that using one outlet is good enough. He's doing this because he knows it will give him a competitive advantage over Boiler Manufacturer #1. Hey, it costs less to pipe a boiler with one outlet than it costs to pipe a boiler with two or more outlets. Which price is going to look better to you, the customer?

So that's what goes into his installation manual.

Contractors begin to buy Boiler Manufacturer #2's boilers over Boiler Manufacturer #1's boilers because #2's boilers are cheaper to install, and because #2's sales brochure says that everything will be okay. Even if it's not as "okay" as it could be. But you've already paid for the boiler, and now you're holding the bag.

At this point, Boiler Manufacturer #1 is getting his brains kicked in because he's being pure while his competitors are stretching this steam-velocity thing to its limits. He throws up his hands and decides that he can't be holier than the church. If everyone else is saying it, then one outlet is probably fine. It's the path of least resistance when it comes to sales; and it seems to be what the customers want (cheap). He's tired of trying to convince contractors that slower steam is drier (and greener) steam. He gives in.

Now the steam leaving that boiler is going to carry more water with it, and that's going to raise the fuel bills for all the years to come.

This goes on.

I think that if there are 10 exit holes in a steam-heating boiler, you're smart to use *all* of them. If you're trying to sell the job to a customer, show that customer these pages. There's no getting around Mother Nature and the laws of physics. High-velocity steam lifts water out of the boiler. The old steam boilers never had holes smaller than three inches in diameter, and there was a reason for that – dry steam.

Rise up!

Okay, the steam leaves the boiler through a hole of a certain size, and that hole will determine its velocity. The steam is now out of the boiler and there's most likely some water in it. That's the nature of modern steam boilers. We're going to use the piping to remove most of that water, so this is definitely *not* the time to reduce the size of the pipe that's in that exit hole. If you do that, you'll increase the steam's velocity at a critical point. It would be like putting your thumb over the end of a garden hose. Please don't do it.

Rise up full size and go as high as you can. The higher you can make this pipe, the drier the steam is going to be. Most boiler manufacturers want you to rise at least 24 inches from the boiler's waterline to the bottom of the steam header (that's the main horizontal pipe you see in the drawing). I think it's better to rise at least 24 inches from the *top* of the boiler, not from the waterline. Again, our goal is dry steam because dry steam saves green.

If the water gets into the system it's going to rob the steam of much of its latent heat. It's also going to create water hammer in the piping, water-level problems at the boilers (the burner will cycle on and off), and you may wind up dumping water. That wastes money because when you dump water you have to add fresh water later on, which may involve chemicals, and chemicals aren't cheap. It will also involve oxygen, which eats boilers.

Rise up as high as you can.

Seeing is believing.

Our friends at Weil-McLain do the industry a great service when they bring their glass-piped steam boiler to the tradeshows. Seeing is believing. Here's a series of photos I took at one of those shows. We begin with a cold boiler. Here's a view from one side.

I can see clearly now!

And here we are on the other side. Nice job with the glass pipe, don't you think?

Okay, we start the burner, make steam, and watch as the water rises up that pipe. Keep in mind this is happening in a perfectly piped boiler. What you're about to see is *normal*. Imagine what happens if you don't pipe that boiler with that rise to the header.

It starts to percolate, and here comes the water

Look how high it is a moment later

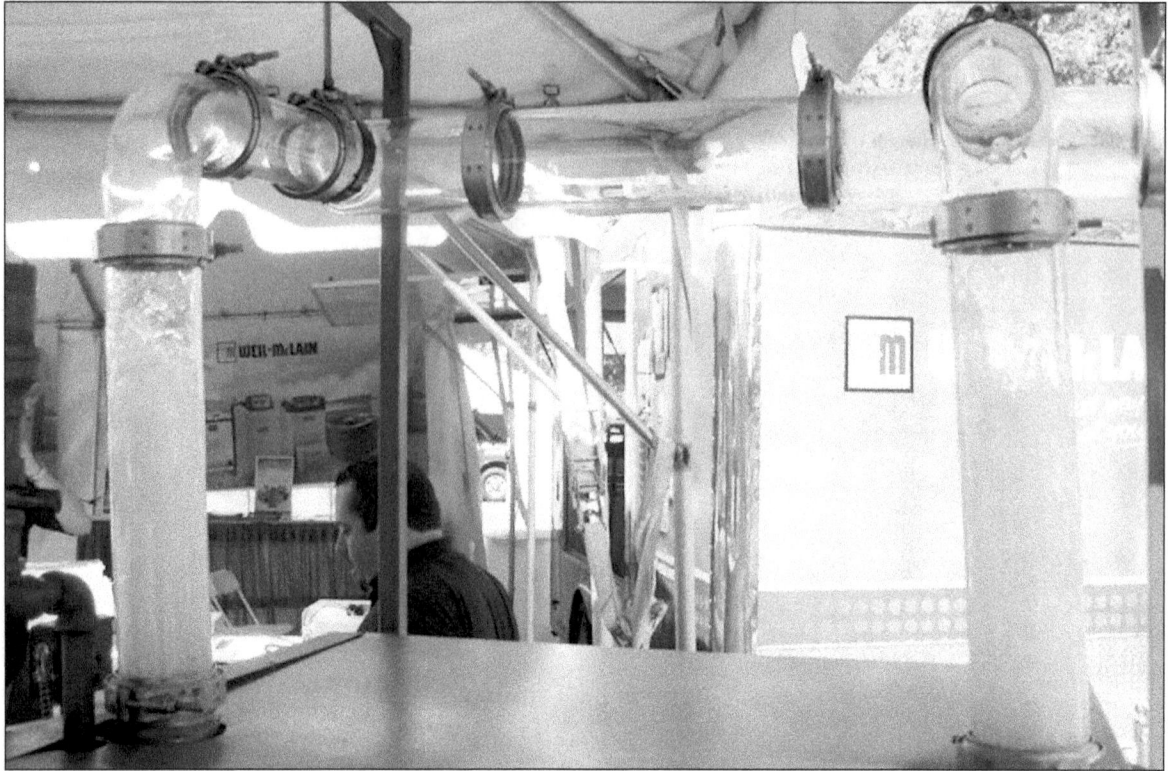

Higher still!

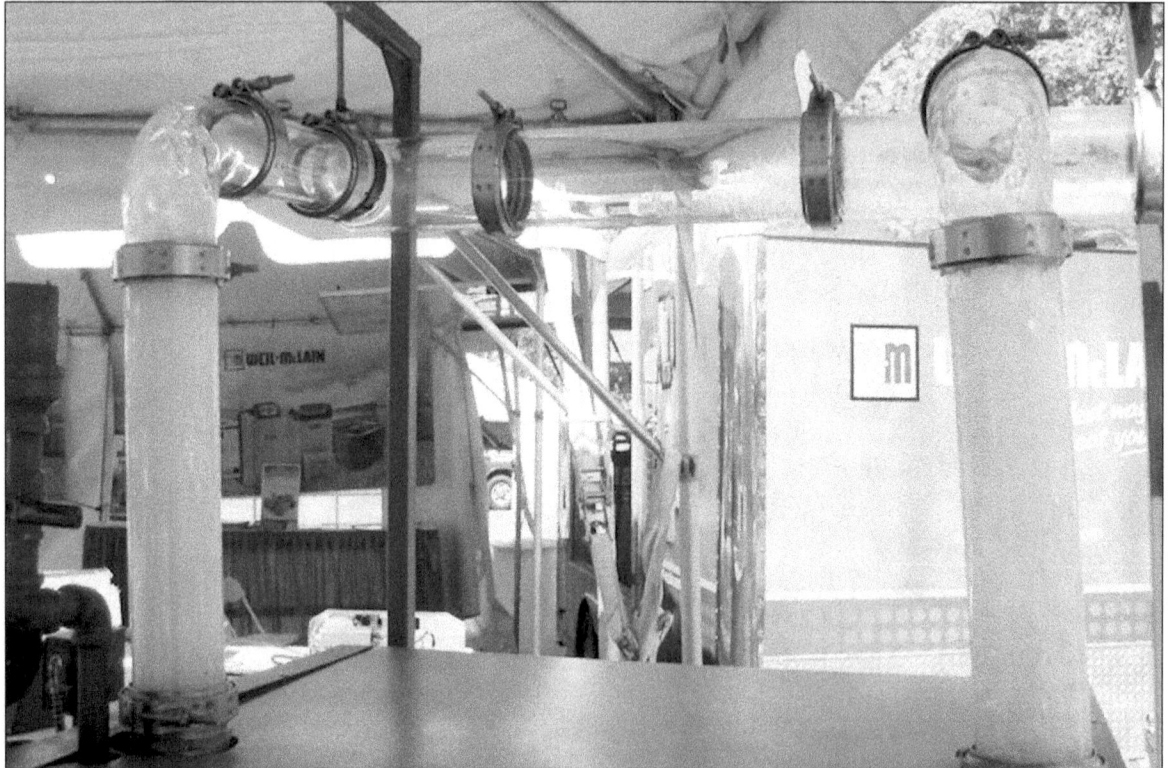

Nearly to the top

Chapter 5: Greening the Piping

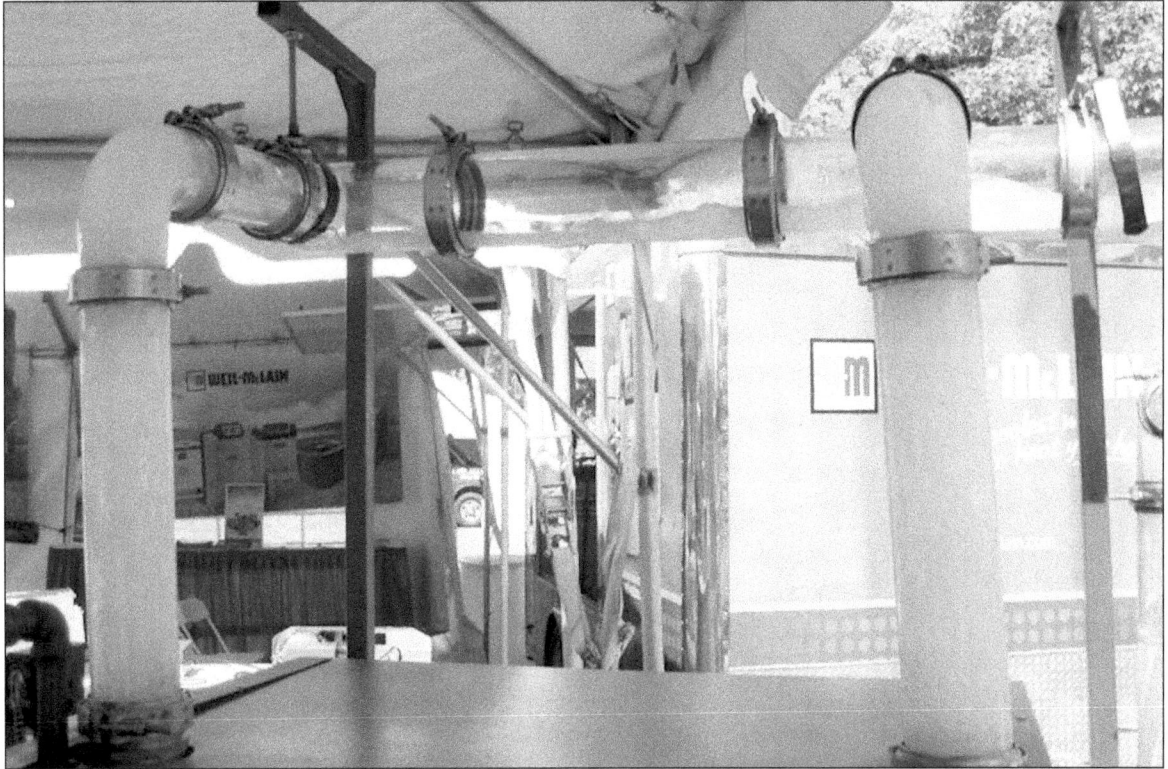

The risers are full and it's spilling into the header

This is normal!

Can you see why you need that 24" rise from the top of the boiler?

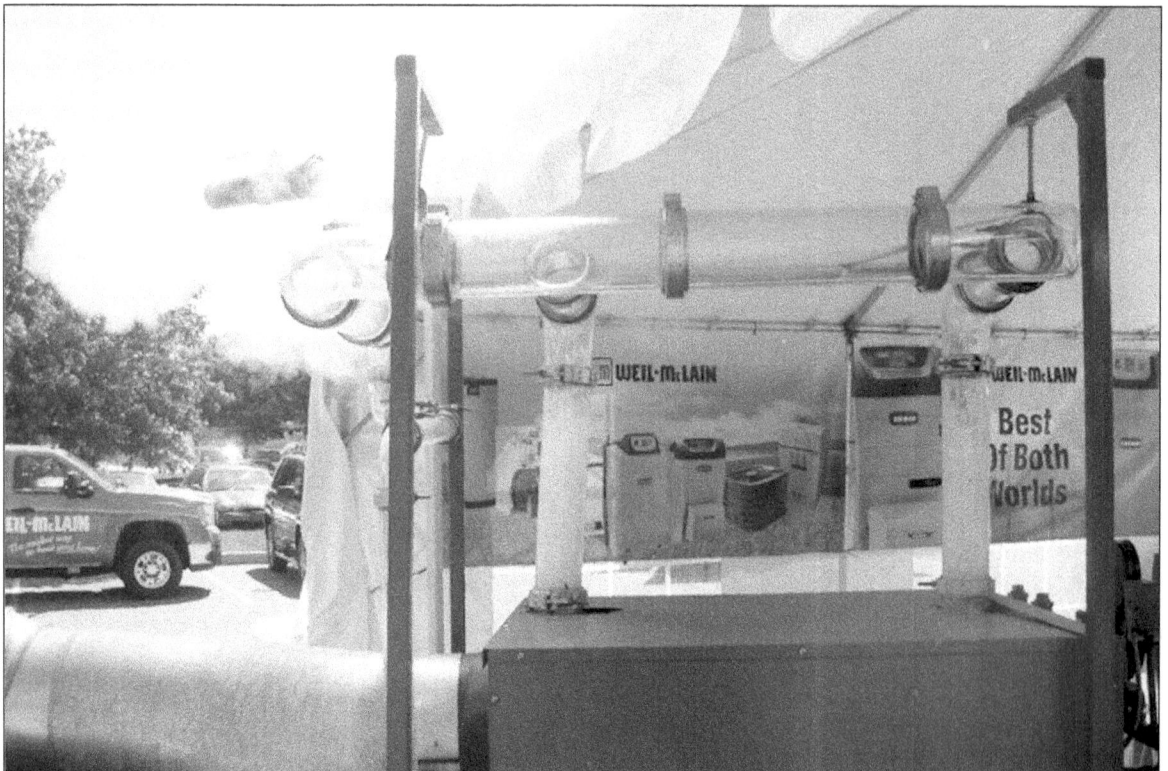

Running for a while now with the valve open

Believe me now? Those 24 inches are critical, and you can see why I'm suggesting that you take it from the *top* of the boiler rather than from the boiler's waterline, regardless of what the manufacturer says. The higher the rise, the drier the steam.

Oh, and if you increase the size of the riser, it's really not going to make much of a difference. The exit hole in the boiler determines the steam's velocity. At most, an increase in pipe size at this point will act only as a small reservoir to contain the rising water. If you want to increase a pipe's size do it at the header. Going bigger here will cost a few more bucks, but it's worth it. Take a look at this photo and you'll see why. The boiler has been running for a while and the steam is drying out.

This is what dry steam looks like

The Header

The header is the main horizontal pipe. Its job is to slow the steam, and, hopefully, the steam will separate from the water at this point. Let's look at that drawing again.

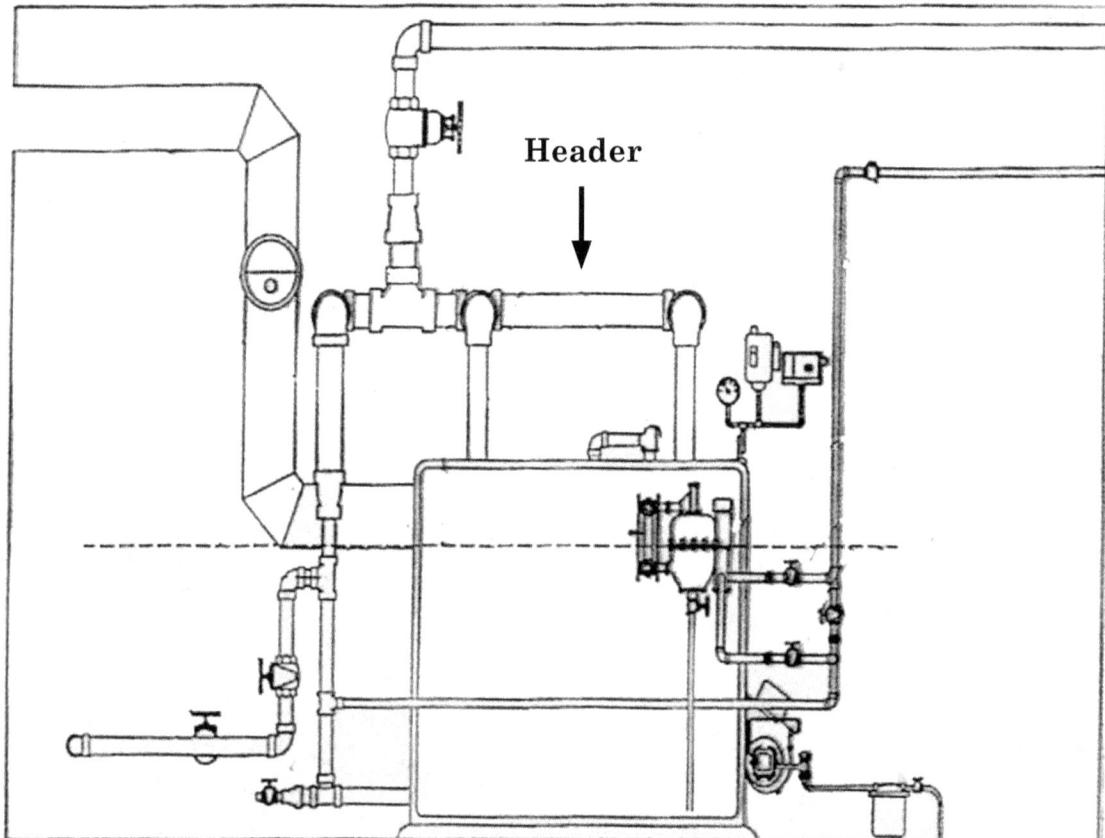

Notice how there's no piping to the system between the risers. The take-off to the system is at a point between the last riser to the header and the equalizer. Pipe this way and you'll get the steam and any carryover water heading in the same direction. You won't have any head-on collisions inside the header that can drive carryover water up into the system. Oh, and be sure to use a tee that's the full size of the header at this point. If you reduce the size of the tee as it connects to the steam mains, the speed of the steam will increase and you'll create a venturi effect that can pull carryover water into the steam mains. The result is wet steam, which wastes fuel.

Notice, too, how the header is offset from the two risers leaving the boiler. This offset allows the header to expand and contract as it heats and cools without putting any strain on the boiler. If that header were directly between those two risers, its expansion would spread the risers in opposite directions, making them act like a crowbars on the boiler sections. This is a great way to break sections. Please don't do it.

The fact that metal expands and contracts when heated and cooled is also the reason why I don't like to see copper tubing used on near-boiler piping. Copper expands a lot more than steel, and we solder copper joints. When they start twisting under the torque of expansion, those joints are going to come apart. Sure, copper is cheaper to install than threaded steel pipe, but a threaded fitting can move a bit with the tremendous forces of expanding metal. It will last longer.

Pay me now or pay me later.

Big steam boilers often have offset headers that are welded in place because big, threaded steam fittings are expensive. Welding is cheaper, but welds won't move. If you're going that route, it's a good idea to install companion flanges in the risers to the welded, offset header. The header will be able to move a bit on those flanges and that will make the boiler happier, and it will last longer.

And before we move on, let me show you what happens when a quick-opening valve opens quickly on a steam boiler that's under pressure. This is what you'll get on those jobs where you decide that adding motorized valves right at the boiler is a fine way to save some green (why heat the whole building when you only use a part of it?). Motorized valves were around in the 1920s, but they didn't put them right near the steam header and here's why.

Mike's about to open that valve. Ready?

Mike opens the valve

All the water in the boiler leaps into the piping...

...and winds up in the parking lot...

...or up in the building where it will hammer like mad

Seeing is believing, right?

Onward.

Drop-header

A drop-header is an old-timers' piping arrangement that takes the risers up high and then turns them downward to connect to the horizontal header. Here's a photo of one.

It's easier to make your pipe connections from multiple risers to a drop header because there's more swing in the pipes. Sure there are a few more fittings and it costs more to do it this way, but you'll have more wiggle room when you're building it, and a drop-header also gives you better angles when you're connecting from the horizontal header to the building's existing steam mains. It also gives your header more places in which to expand and contract, so that's better for the boiler.

Drop-header

Another nice advantage of the drop-header is it gives the risers leading to the header more height. Check it out in the photo. You're making the steam climb higher before turning to enter the header, and that will leave more carryover water behind. Go back and take another look at that sequence of photos I showed you of the glass-piped Weil-McLain boiler and you'll appreciate the difference riser height can make when it comes to dry steam. The higher the better.

When you make the turn and drop down to the header, you can go as low as you'd like with the header as long as you keep it above the top of the boiler. The 24" riser height that the boiler manufacturers want goes from the boiler's waterline (or, as I prefer, from the top of the boiler) to the top of the risers before they turn to connect to the drop-header.

And you can use a drop-header with a single riser as well as with multiple risers to the header. They cost more, but I think they save installation time, and they do make a difference in the quality of the steam, which saves green.

The Equalizer

At the end of the horizontal header we have a pipe that turns downward and connects to the bottom of the boiler. This is the equalizer and it gives the carryover water a route back to the boiler. Take another look.

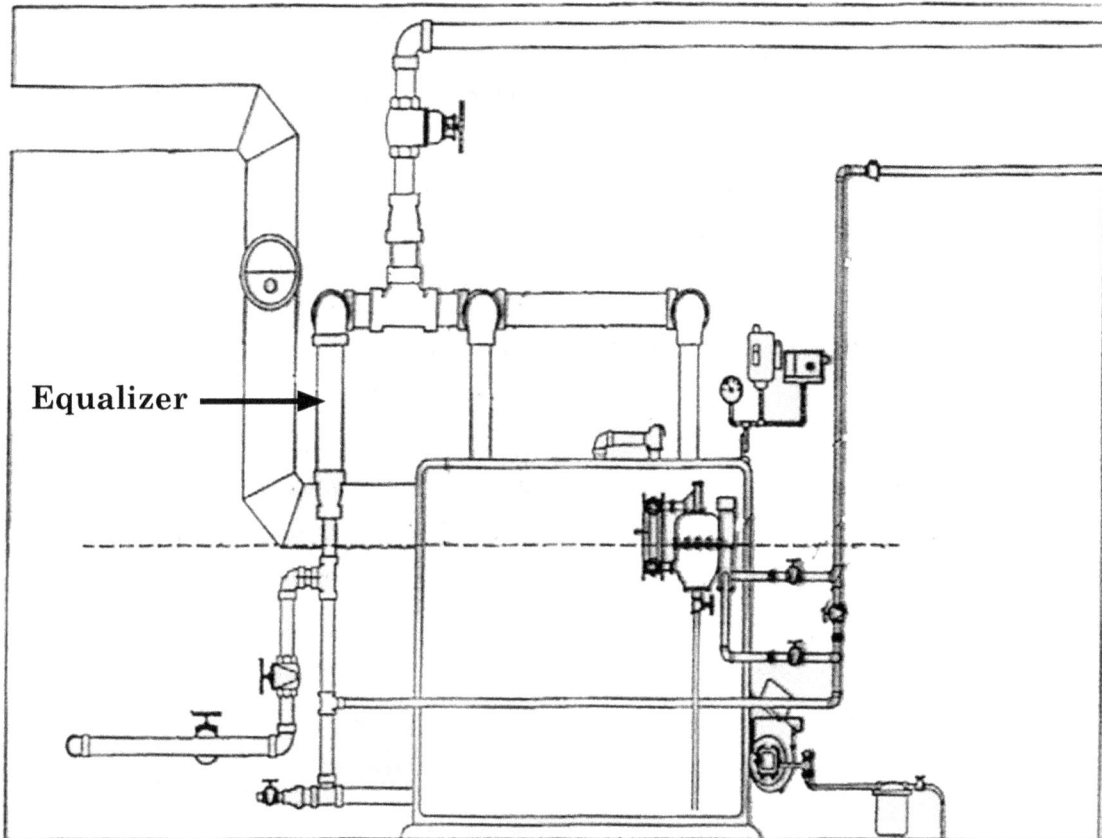

An equalizer also puts just enough steam pressure on the return side of the boiler to keep the water that's inside the boiler from backing out as pressure builds (that's where the name comes from). And because you don't want to lose any of that pressure to piping pressure drop (caused by friction as the steam zips through it), you have to use a certain size equalizer for a certain size boiler. Here are the guidelines from the 1930s:

Gross boiler rating in Sq. Ft. EDR	Size of the equalizer
Up to 900	1-1/2"
900 to 6,400	2-1/2"
Over 6,400	4"

Note how they jump from 1-1/2" right to 2-1/2", skipping over 2". Their goal was to keep the friction loss of the steam to a minimum. These sizes worked then, and they still work now. Physics doesn't change. And please don't ever use an equalizer smaller than 1-1/2". You'll have problems keeping the water in the boiler.

If the system is using a condensate- or boiler-feed pump to put the water back into the boiler, the equalizer is not going to equalize anything because the pump's receiver is open to the atmosphere, and the pump has a check valve at its discharge. The water can't back out of the boiler as long as that check valve is working, so the equalizer now becomes a header-drip. Its only job is to put the carryover water back into the boiler. An equalizer is only equalizing pressure between the supply and return side of a boiler when it's on a gravity-return system.

Which brings up the question: Should you use a check valve on the return of a gravity-return system to avoid having to use an equalizer? The answer is no, and for a couple of good reasons. First, a check valve will get stuck open because there's often debris in the return piping. The equalizer does the job of a check valve, and it's not going to let debris affect its operation. That's how it came to be. There were too many problems with check valves in the early days of steam heating.

The other reason a check valve is a bad idea is because, without the equalizer, you have no way of getting the carryover water back into the boiler, which leads to wet steam, and wet steam leads to high fuel bills.

The Hartford Loop

No, we're not talking about the highways that loop around Hartford, Connecticut; we're talking about The Hartford Steam Boiler Insurance and Inspection Company. They're named for that New England city, and in 1919 they mandated a return-piping configuration that would help prevent boiler accidents. The trade gave this piping its unique name in honor of that fine old company.

Before 1919, installers would pipe boilers like this.

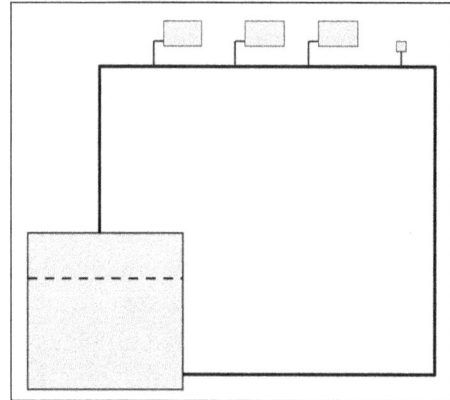

If a return line broke on a coal-fired boiler, the water in the boiler would quickly flow out, and that could cause a fire. There were no low-water cutoffs for coal-fired boilers. And if some poor soul came along and added water to that red-hot boiler, it just might explode and take down most of the building. That happened all too often in those days.

Hartford Insurance insured boilers and they needed a way to lessen the risk of dry-firing and boiler explosions, so they came up with this piping arrangement. Notice how the return pipe connects to the equalizer at a point just below the boiler's waterline.

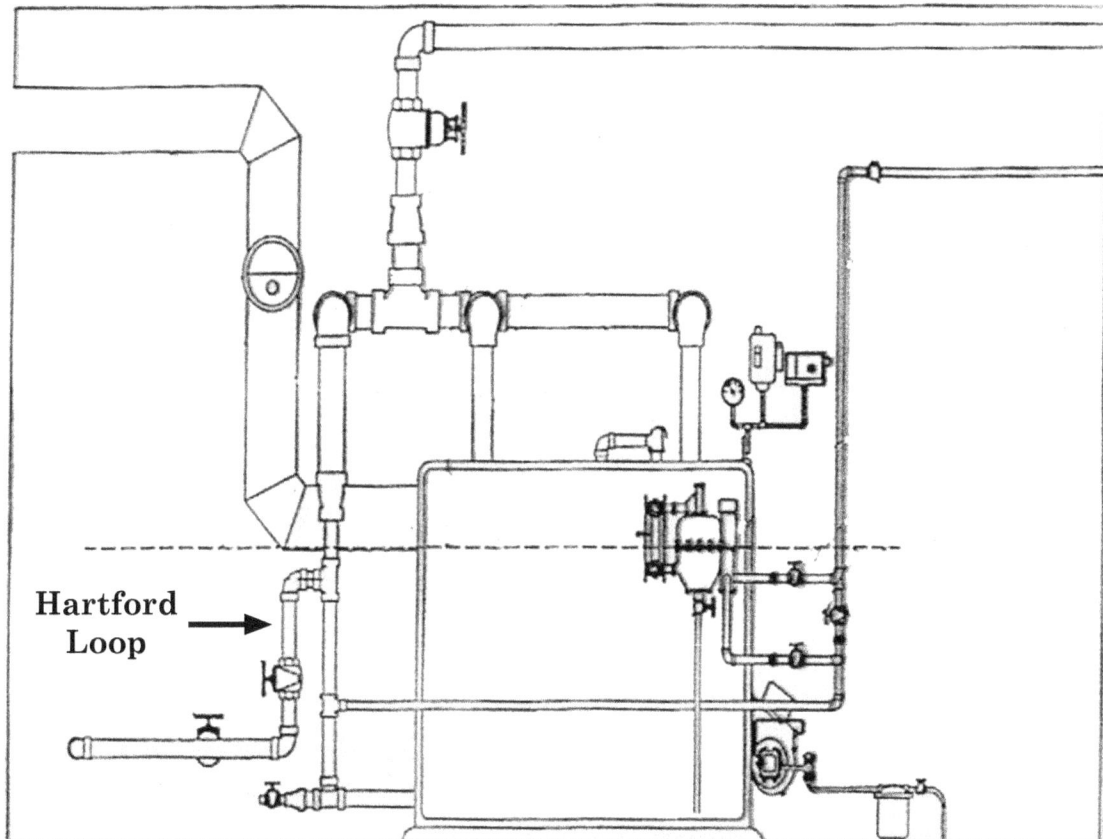

Hartford Loop

If a return line should break, the water in the boiler will still flow out, but only to the point where the Hartford Loop's close nipple connects to the equalizer. It bought you some time in a coal-fired era. And even though you're now firing oil or natural gas, the Hartford Loop is still a great thing to have working for you because it backs-up the low-water cutoff. It's very green because it can save a boiler. It will also make the boiler inspector smile when he comes to check out that installation.

I mentioned the *close* nipple and it's important that it be a close nipple because the returning condensate is relatively cool, compared to the water that's inside the boiler. That water has steam in it. The steam is rising through the equalizer because the equalizer connects to the bottom of the boiler. When the return water meets those rising steam bubbles it will cause those steam bubbles to quickly turn back into water. That leaves these tiny "holes" in the boiler water and the return water will slam into those holes because Mother Nature hates a vacuum. That slamming can result in water hammer right at the point where the return pipe connects to the equalizer. By using a close nipple, you shorten the distance the returning water has to travel as it makes the turn to enter the equalizer, and this lessens the potential for water hammer. If you use a long nipple at this point, it's like giving the returning water a runway to work with. The water will fly down that runway and slam into the back of the tee that connects to the equalizer. Water hammer isn't pleasant, and it's not at all green.

King valve and return valve

Here are a couple of valves that work well together.

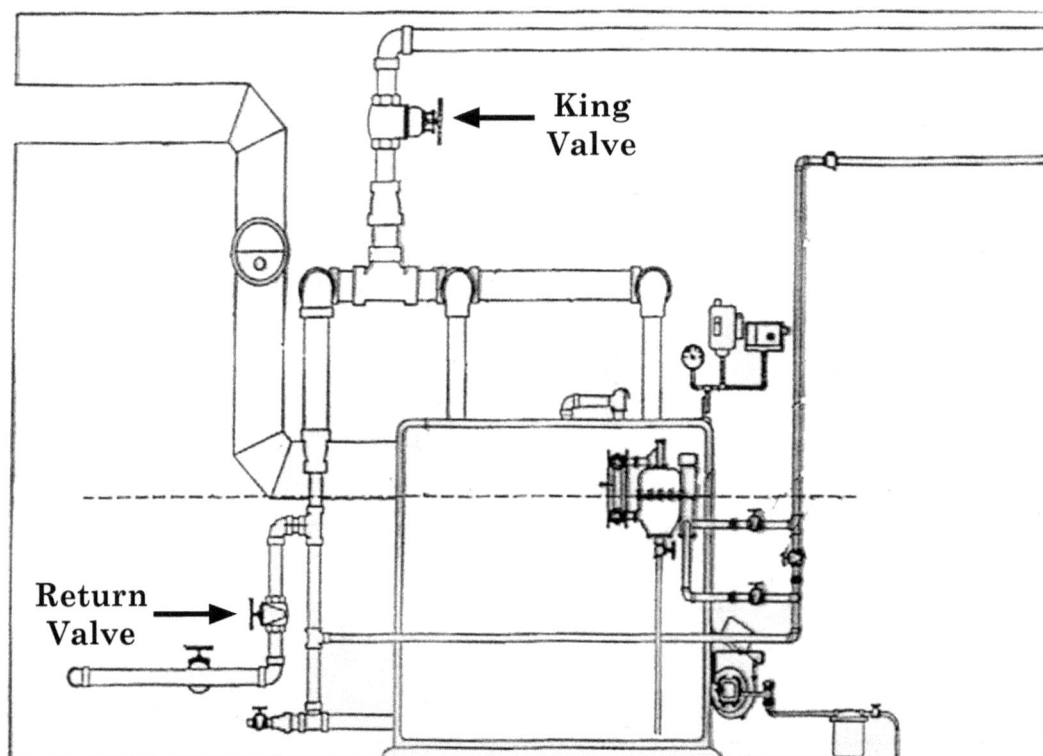

The king valve is the one on the main steam line leaving the boiler. It gives you a way to make steam without sending it out to the building. The return valve is the one on the return line (great name for it, eh?). It's just before the Hartford Loop.

All steam boilers fill with goop as time goes by because all steam heating systems are open to the atmosphere. Pipes corrode over time, and solids in the feed water remain behind as the steam forms and leaves the boiler. All of this settles to the bottom of the boiler and winds up in the mud leg. If you don't get rid of the goop, it will slow the water's circulation through the boiler, creating hot spots, which will shorten the life of the boiler.

The boiler winds up in a landfill and you wind up with less green in your pocket.

Used together, the king valve and return valve give you a way to isolate the boiler so that you can blow-down the mud leg. I think you should do this at least once a year. Filling the boiler with water and trying to use the static weight of the water in the boiler won't get the job done. A column of water 28-inches high exerts just 1-psi pressure at its base. There's just not enough water in a steam boiler to give you a good flush with gravity alone. You need some pressure; so close those two valves and build up about 10-psi steam pressure. It won't take long to do that because you're able to isolate the boiler from the system. Be careful not to get the pressure close to the 15-psi setting of the relief valve because when that pops it's a maturing experience, and one you should always avoid.

When you reach the 10-psi pressure, open the mud-leg's valve (it works best if this is a full-size valve). You can see that valve in the drawing (sadly, it's not full-size). Open the valve, stand back and be careful. What comes flying out is hot stuff.

Return valve and drain for wet return

These two valves are down at the bottom. One we already looked at; it's the return valve. You used it with the king valve to blow down the mud leg. The other is a drain valve for the *system* side of the return valve. Take another look at the drawing. See it there on the horizontal return, just before the return valve?

Use these two valves to waste the condensate when you first start a new boiler. A new steam boiler often acts like a scouring pad on an old steam system. You get a lot of corrosion and scale flowing back to the new boiler, and it's better to catch that stuff before it has a chance to enter the new boiler and contaminate the water, or worse.

As I mentioned earlier, MEX is a good soap for system cleaning. You use one pound per 50 gallons of boiler water. Pour it in through the top of the boiler, perhaps through that funnel you made from fittings. Let it percolate through the system and catch it all on the return side with those two valves. It works like a charm, but be sure to get rid of all the MEX once it's done its job because that soapy foam can create wet steam if you leave it in the boiler.

A clean steam boiler is a green steam boiler.

Pressuretrol and gauge location

Let's take another look at our near-boiler piping. I really like this drawing because it's practically perfect. But there is something I'd do differently and I've highlighted it here.

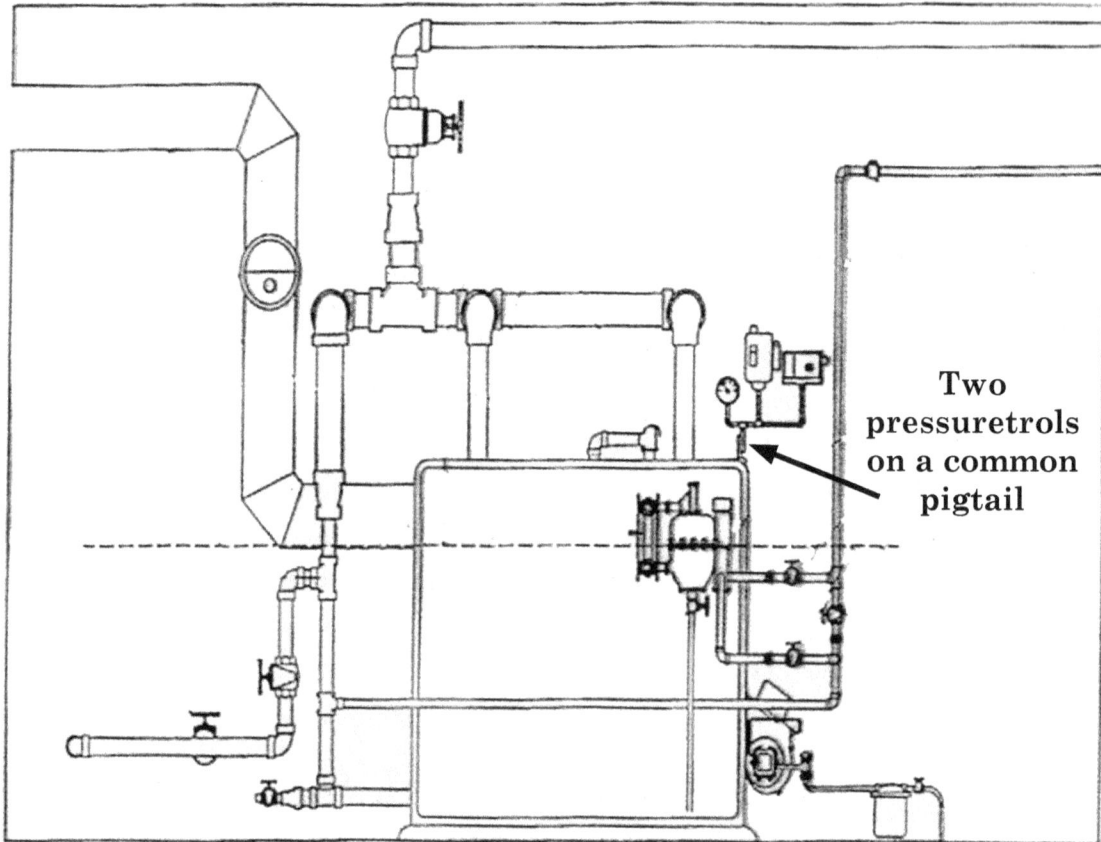

Two pressuretrols on a common pigtail

This boiler is going into an apartment building so it has two pressuretrols. One operates the burner, starting and stopping it at a low- and a high-pressure setting. I told you about how all of this works earlier. The second pressuretrol has only one setting and that's OFF. We'll set that at 10-psi and hope the pressure never gets that high. And if it does, someone is going to have to manually reset this pressuretrol because it's a safety device. A low-pressure boiler in a space-heating application should be operating at a low pressure (keep the Empire State Building in mind). If it goes to 10-psi, then something is wrong. If it goes any higher, it's probably going to pop the relief valve, and if people are in the boiler room when that happens they are going to have that maturing experience I mentioned earlier.

By the way, we use *two* pressuretrols in commercial- and apartment buildings because there are usually lots of people in those buildings and we're trying not to blow them up. That's one of the things that's always made me wonder. Are the people in a single-family home less valuable than the people in an apartment building or an office building? Why don't we go to this extent to protect *them*?

I think that's a good question to ask, don't you?

In a commercial- or apartment building one pressuretrol is watching the other pressuretrol. It's like a big brother looking out for a little brother, and both brothers are supposed to be looking out for you because you're one of many people in that building. They've got you covered.

Now take a look at how the two pressuretrols are piped. See how they sit atop that one pigtail? This presents a potential problem.

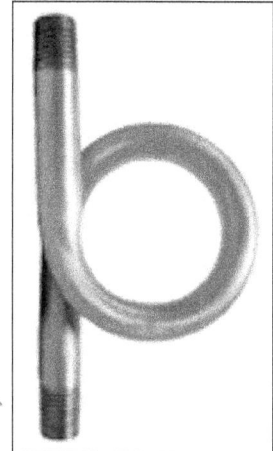

Pigtail

But wait a minute; you may not know what a pigtail is. Here's a photo of one.

Good name for it, don't you think? We call it what it looks like and it has two functions. The first, and most important, is to fill with water and keep the latent heat of the steam out of the control. This gives more life to any control or gauge. These devices are sensing *pressure*; they don't need to feel the full heat of the steam to do that.

A pigtail's other function is to clog. It sets out to do this from the first day. Just look at it. There's one way in and no way out for any debris that's on the surface of the boiler water. Anything roiling around and bouncing off the surface of the water has a good chance of winding up in the pigtail. It's like shooting baskets with a net that's sewn closed at the bottom. Put enough debris into that pigtail and it will clog solid, and when that happens, the pressuretrol won't sense the pressure in the boiler. That pressure will continue to rise and the second, manual-reset pressuretrol will have to do its job of protecting its little brother.

But look at how we've piped the two pressuretrols. They're both on the same pigtail. So when the pigtail clogs it takes them both down. That relief valve is bound to pop, which is never pleasant with steam.

So pipe *one* pressuretrol or *one* gauge to *one* pigtail. It's the green thing to do because relief valves that pop cost money.

Low-water cutoff

Okay, one more thing, and it's a *very* important thing – the low-water cutoff. Its job is to stop the burner if the water level in the boiler goes too low. If you fire a boiler that doesn't contain enough water, the boiler will overheat to a point where it can burn down the building, or even explode if someone comes along and adds water while it's that hot.

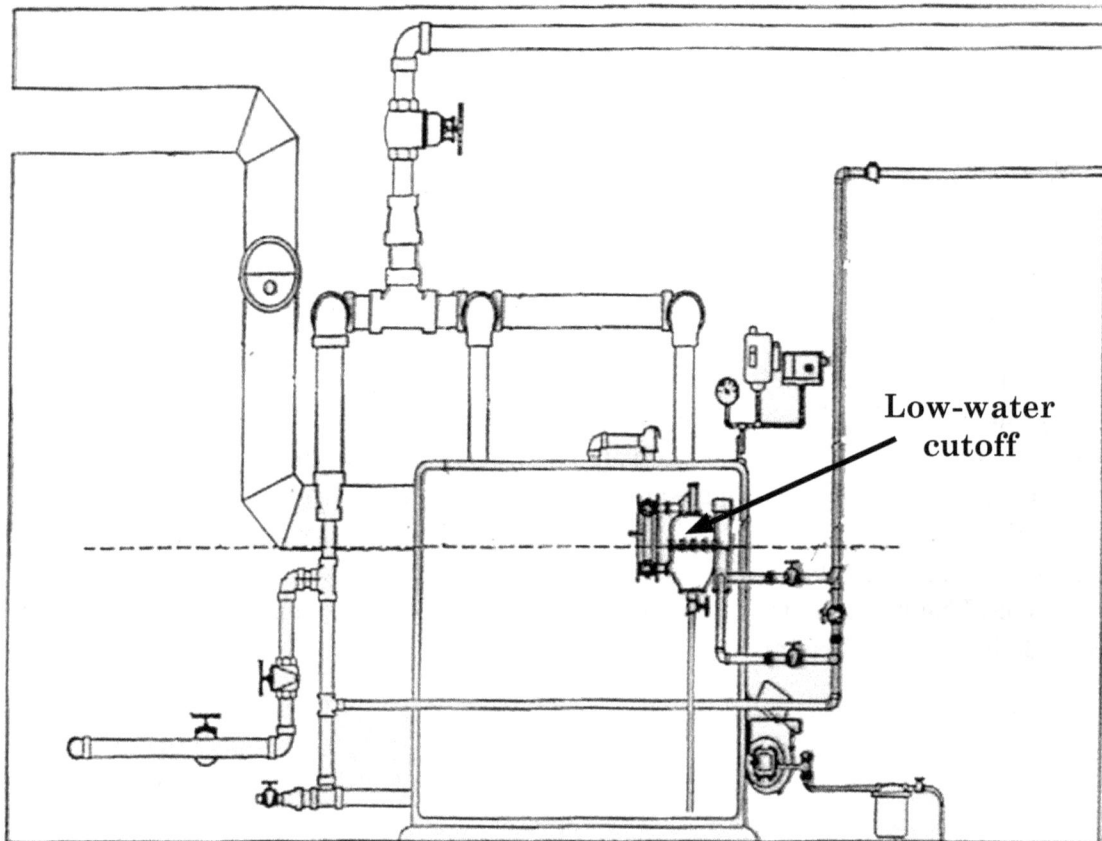

Low-water
cutoff

The low-water cutoff will protect against a low-water condition, and the one you see on this boiler will also automatically feed water into the boiler to maintain a safe, minimum waterline. It does this with a mechanical float that's similar to the automatic valve in a toilet tank. If the float drops too low, it trips an electrical switch that's wired in series with the burner.

The manufacturer of the low-water cutoff/automatic feeder states that someone has to blow down the float chamber on this unit once a week. If there's no one around to do this, the instructions still apply. Mother Nature doesn't care about a lack of staff. She will continue to shove goop into that float chamber, and if enough goop gets in there, it will look like this.

As you can see, the float is going to have a tough time moving up and down when there's that much goop in the float chamber. The result is a dry-fired boiler, which is exactly what happened in this case.

If you have low-water cutoffs with mechanical floats you *must* blow them down every week. There's no way around this.

There are other types of low-water cutoffs that work with electronic probes instead of mechanical floats. These sense continuity through the water and shut off the burner

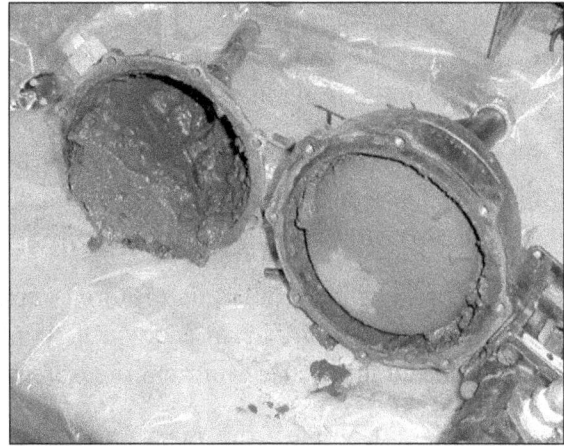

Clogged float chamber

if there's no water present. Using these gets you away from having to blow-down the float chamber once a week, but you should remove any probe-type low-water cutoff at least once a year to clean and inspect it.

To me, the green thing to do is to use two low-water cutoffs on every steam boiler. Make one mechanical and the other electronic and place them in different spots on the boiler. It's like wearing a belt and suspenders and it's a wise investment in safety. Low-water cutoffs are cheaper than boilers, and they sure are green.

Now let's move off the boiler and out into the system.

Supply piping

How can you green those horizontal pipes that carry steam from the boiler to the vertical risers that reach up to the radiators? Well, the first thing to do is to check the pitch. If the steam and condensate are traveling in the same direction (we call this parallel flow), the pitch should be at least one inch of downward slope for every 20 feet of travel. If the condensate is flowing backwards against the onrushing steam, the pitch needs to be double that – one inch in 10 feet. If the pipes sag at any point, condensate will gather on the down cycle, like puddles in a poorly paved road. When the steam returns, flying down the pipe at 60 MPH or faster, it will pick up those puddles and shoot them down to the end of the pipe. The water will hammer against the fittings and rattle the building. Water hammer is destructive and that's not green.

How do pipes lose their pitch? Usually it's people doing it. They hang stuff on the pipes. I once visited a house where the pipes where doing the mechanical mambo; this just started to happen that winter. Turns out the homeowner had hung one of those sand-filled heavy bags on his steam main. It was great for his karate practice, but not so good for the steam. The condensate was hitting the elbows harder than that guy was hitting the bag.

If a steam system ever floods because someone left the water feed valve open for too long, the weight of the water in the steam pipes can cause the pipe hangers to give, and that will create sags in the horizontal supply lines, inviting water hammer.

What else can you do to green supply mains? Insulate them. Insulate anything that's not a radiator. Tuck in the steam and it will have a better chance of making it to where the people are. That saves money.

Don't add more load (radiators) to a supply main than the supply main can handle at the design pressure drop. If you do, you'll wind up with either radiators that heat unevenly, or needing to run higher-than-normal pressure on the system. Both cost money.

For example, here's the maximum acceptable capacity for a horizontal, parallel supply main in a one-pipe-steam system.

Pipe Size	Maximum Square Feet EDR
2"	386
2-1/2"	635
3"	1,163
4"	2,457
5"	4,546
6"	7,642

That chart is from my book, *The Golden Rules of Hydronic Heating*, which you can get in the **Shop** at HeatingHelp.com. The book contains a chart like that for every pipe in both one- and two-pipe steam systems. When you know the limits to what a pipe can handle, you can save money.

If it's serving one-pipe steam, a supply main should have a main vent near its end, but not right *at* its end. Ideally, it should be 15 inches back from the elbow that turns down to send condensate back to the boiler, and at least six inches up on a nipple. Piping the vent this way protects it from potential water hammer and adds years to its life. We'll talk more about this when we get to the Greening the Venting chapter, which is coming up next.

The Risers

If the steam isn't getting to the people, it's not green. They'll complain and be miserable, or buy electric space heaters, or move out.

My daughter Meghan lived in an apartment building in New York City. Like so many NYC buildings, this place had one-pipe steam. Meg lived on the fifth floor, which was the highest in the building. The steam had a long way to climb to get to her radiators

In Meg's kitchen there was a bare pipe that came out of the floor and ended near the ceiling. There was another just like it in her bathroom. These were the steam risers, and they had no insulation on them because they were also the radiators for those rooms. This is typical for older apartment buildings. Here's what they do:

Riser size	Square Feet EDR per Linear Foot of Vertical Rise
1-1/4"	.63
1-1/2"	.73
2"	.88

For instance, if there's a 2" bare riser going up through a bathroom and the ceiling is eight-feet high, that pipe will give off .88 EDR per foot, for a total of 7.04 EDR. One square foot EDR is 240 Btuh with steam, so we have a total of 1,689.6 Btuh coming off that riser. That will get the job done in a small bathroom or kitchen.

Meg called me on the first cold day she was in that apartment.

"Dad, there's a noise coming from that little silver thingy at the top of the pipe in my living room," she said. (I immediately disowned her because any daughter of mine should know the proper names of all the components in a steam system. It's only right. Silver thingy, indeed!)

"Do you mean the air vent, Meg?"

"I don't know; it's silver."

"Does it look like a bullet?"

"Yes."

"That's the air vent, and it's the wrong size."

"What should I do? It's making a lot of noise."

"Well, you can tell the super, but he's probably going to tell you to just get used to it."

How's that for green advice?

This riser was an insulated pipe. It was delivering steam to the radiators in all the living rooms on that vertical line. It wasn't a radiator itself, as the pipe in Meg's kitchen and bathroom were radiators. And since Meg lived on the top floor, she had the air vent for the whole riser entering her apartment. When the steam came up, it had to push out all the air in about 50 feet of 2" pipe. That's a lot of air. The vent at the top of riser was a typical radiator vent, sized to release the amount of air you'd find in a relatively small radiator. It was struggling to deal with all the air in that riser, and that's why it was making noise. Blow hard through a small hole and you'll get noise.

I knew that if Meg called the superintendent he would, at best, replace the too-small air vent with another too-small air vent and the problem would remain. It would be difficult for the super to install a vent with greater capacity because there was a 1/8" hole drilled in the pipe, and that hole would accept only a radiator vent. This is typical of older apartment buildings. The right air vent would allow the air to escape faster, which would allow the steam to move faster. That would save fuel because the radiators would get hot faster. But it costs a bit of money to save a *lot* of money, and that's why cities like New York are filled with people like Meg who have to listen to hissing air vents as they sit shivering in their apartments.

It *is* easy to be green, but you have to think things through, see the steam and the air in your mind's eye, *and* be willing to invest a few bucks to save a *lot* of bucks.

More on this in the next chapter.

But before we leave the risers behind, I need to gently suggest that you not hang holiday lights on them. I say that because a friend recently moved to NYC from California and did just that. She wrapped strings of lights around all her steam risers, both the insulated- and the uninsulated ones. The lights didn't last long. There's this thing about electric wires and ambient temperatures. You can read all about it on that tag that's wrapped around the wire near where it plugs into the wall. Mother Nature doesn't care that it's the holidays. She'll burn down your building anyway.

Which brings up another good question: Can steam risers set paper on fire? Well, the surface temperature of any pipe in a steam-heating system is supposed to be no more than 215-degrees. The kindling point of paper is 451-degrees Fahrenheit (remember the Ray Bradbury book, *Fahrenheit 451?*). But it's still not a good idea to tape your holiday cards to your steam risers. The glue gets sloppy.

Just leave the risers alone and let them do their job. They're pretty enough as is.

Piping at the radiators

We now have the steam up in the rooms with the people. If the piping is wrong at this point, the people are either going to open the windows (if it's too hot), or bring in electric space heaters, or run the oven with its door open (if it's too cold). Not very green either way.

If you're working with two-pipe steam radiators, the steam is entering the radiator through either the top or the bottom. Steam is lighter than air so it will always head for the top of the radiator. It's okay to throttle the supply valve on a two-pipe radiator because the condensate doesn't need to share the space inside that valve.

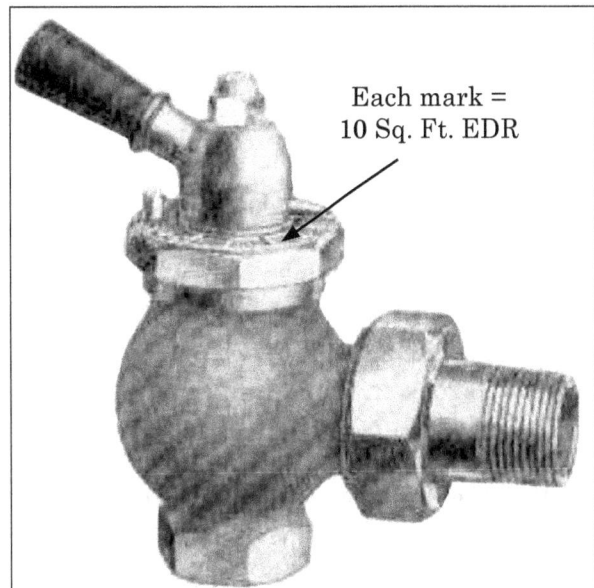

Each mark = 10 Sq. Ft. EDR

Supply valve with variable orifice

Some two-pipe systems may have an orifice inside that supply valve. Its job is to limit the amount of steam that can enter the radiator. An orifice will typically allow in about 80% of what the radiator can condense. This is great because it can add years to the life of the radiator's thermostatic steam trap. That trap doesn't have to work as hard; it's seeing mostly condensate. Some of the Dead Men made their own orifices by cutting circles from Prince Albert Tobacco tins and then punching a hole through the circle with a nail. If you want an orifice that's more precise than that, contact the good people at www.tunstall-inc.com. They'll size the orifices for you, and from there, it's as simple as slipping these inside the supply valves at the union connection.

Some supply valves had a variable orifice, which looked like the one in the picture above.

See those gradations on the valve's bonnet? Each represents 10 Square Feet E.D.R. By moving that pin, you can limit the swing of the supply valve's handle to 80% (or whatever) of what the radiator can handle. The Dead Men often oversized the radiator for the room's heat loss, and then limited its output with valves such as this one. The goal was to provide comfort, while increasing the life of the steam trap.

There may be an orifice in the return side of the radiator. Many vapor systems had them. And, by the way, a vapor system was any steam system that operated with a pressure of eight ounces or less, so use a vaporstat on these systems. Turn the page to see an example of an orifice on a return elbow.

This one is from Trane. From the outside, it looks like an ordinary return elbow, but there are two small holes on the inside. The upper one allows air to pass; the lower one causes condensate to back up, and that condensate stops the steam from rushing into the return. This particular orifice return elbow has caused some interesting problems for people who have tried to convert a

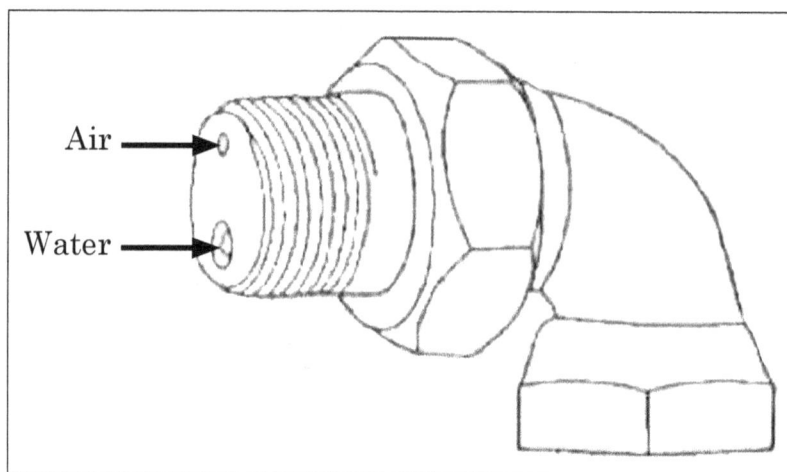

Return-side orifice

system that has these to run on hot water. Go ahead; try pumping water through those tiny holes. You get some interesting velocity noises from them. Every dog in the neighborhood will notice.

If the radiator has a steam trap, make sure it's working. We'll talk more about this in the chapter on Greening the Traps.

Check the pitch of the pipe that leads to the radiator and make sure there's no place for condensate to gather between heating cycles. This will cause water hammer on start-up.

If you have one-pipe steam, the supply valve will always be at the bottom of the radiator. You can't throttle this valve because the steam and the condensate have to share this confined space. The valve has to be either fully open or fully closed. If the radiator doesn't heat, check the supply valve for loose parts by opening its bonnet. Since these valves are rarely opened or closed, the valve seats have a way of rotting and falling into the opening. If the radiator is on the top floor of the building, with the valve connected right at the top of the riser, that rotted-out valve seat can fall into the riser and wind up at the bottom of that pipe down in the basement. This will slow (or stop) the steam in the entire riser.

Check the pitch of those one-pipe radiators. It should slope back toward the supply valve. You need a pitch just of one inch in 10 feet, which isn't much. Usually all it takes is a couple of checkers under the two feet of the radiator, opposite the supply valve. And here, you get a choice of either red or black (or both!).

Never run a supply pipe through one radiator and then on to the next radiator. The condensate that forms in the first radiator will keep steam from reaching the second radiator.

Return piping

If the water can't get back to the boiler, the boiler won't run. A steam boiler that doesn't run will save fuel for sure. In fact, it will be 100% efficient, but the folks in the building won't like it very much.

If the water's not going back to the boiler, then where is it going? It has to go someplace, right? I figure it's either staying in the piping or it's leaking from the system. If it stays in the piping it's eventually (or quickly) going to cause water hammer, which is annoying, destructive and not green. If it leaks from the system, you have to replace it, which costs money. Not green!

If the system has an automatic water feeder, the feeder will do its job by doing what its name suggests – to feed water automatically. But if the water is hung up in the system because the piping is partially or fully clogged, that water may eventually work its way back to the boiler, which will then flood. At that point, you'll have very wet steam flying out to the system, which makes things worse. Someone has to dump water, which costs money.

There are two types of return piping. First, there's the wet return. This is any pipe below the boiler's waterline on a gravity return system (that would be any system that doesn't have a condensate- or boiler-feed pump). I assume all wet returns are clogged, and if they go under the concrete floor at any point, I assume they're leaking. As far as I'm concerned, old wet returns are guilty until proven innocent.

Think about how slowly the condensate moves through a wet return. It's completely filled. A drop of water enters near the end of the steam main. At the same moment, another drop of water will flow into the boiler at the other end of the return main. It's a big U-tube. Drip, drop. That's how it works. Wet returns are constantly filled with acidic condensate, which is slowly munching its way through the steel. If the pipe is underground, it's hidden and no one ever notices the corrosion. That's the problem.

The wet return is also a low place in the system, and a welcome nest for any goop that wants to gather there. And it will. You have to flush wet returns from time to time, but this is a messy job and usually overlooked, especially if no one provided valves for flushing.

Guilty until proven innocent.

The other type of return is the dry return, and this is any pipe that's carrying condensate and is above the boiler's waterline. These last longer than wet returns because condensate doesn't spend as much time inside of them, and they don't run under the basement floor.

So now we have the water back at the boiler. How are we going to get it *into* the boiler? The boiler's under greater pressure than what we have in the return lines. Low pressure won't go to high pressure, so we're going to have to give it some help. And that brings us to this:

The "A" Dimension

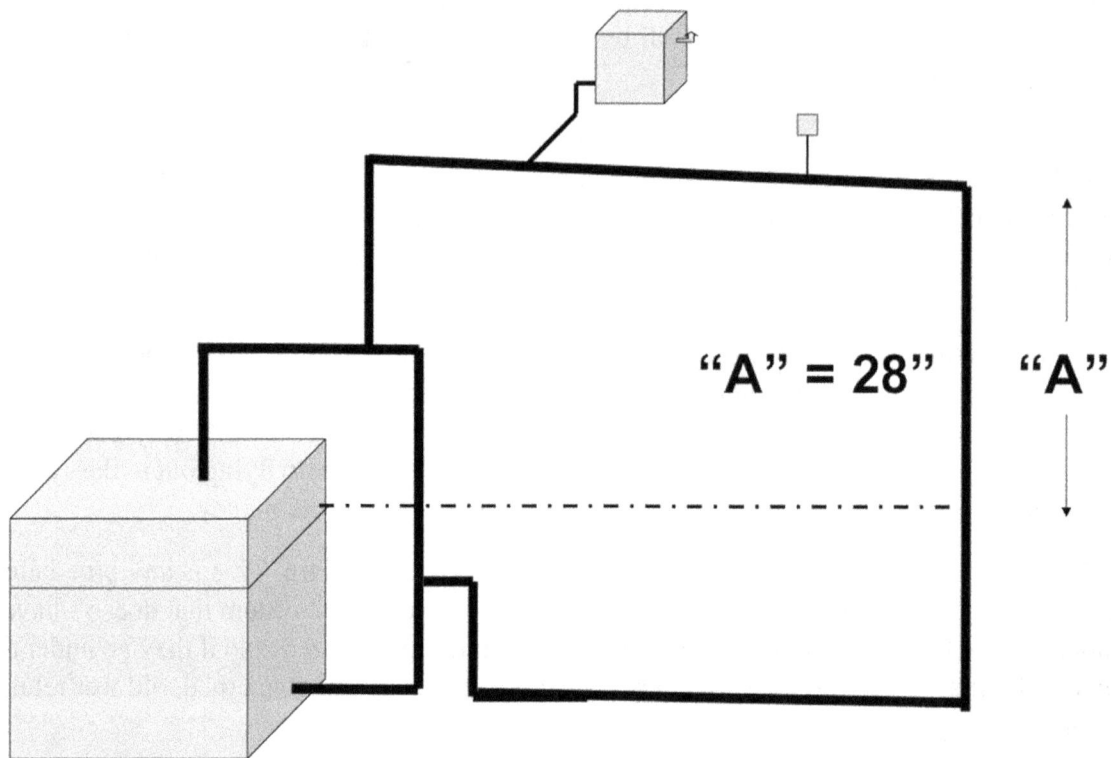

"A" = 28"   "A"

Think like steam. You're inside the boiler and you're trying to get out. The only thing standing in your way is air and the friction that the piping will cause as you race through it.

You leave the boiler and head for the ends of the mains and the radiators. Along the way, you lose energy and you just don't have the push at the end of the steam main that you had back at the boiler. You're tired.

Look at the drawing. See how the wet return forms a U-tube? There's boiler pressure on the left side of that U, and less pressure on the right side of the U. The water wants to stack up on the right because of this, so let's give it some room.

A column of water that's 28 inches high will exert one pound per square inch of pressure at its base. That's what we're going to allow for that dimension marked "A". It's the vertical distance between the boiler's waterline (let's say the center of the gauge glass) and the bottom of the lowest steam-carrying pipe, wherever that might be in the building. The weight of the water that stacks in the "A" Dimension combines with the leftover steam pressure inside the pipe at the end of the main to overcome the pressure that's inside the boiler. The water returns to the boiler.

But what if that wet return line is clogged with goop? If you were the water, you'd have a tough time pushing through, wouldn't you? You'd probably back up into the end of the horizontal steam main, right?

That's what happens with a clogged return. The condensate gets into the pipe with the steam, and the steam starts banging it around. You hear it as water hammer, and it happens in the middle of the firing cycle. If there's a main air vent on that dry return line, the water hammer will probably wreck it, and when main vents go down, fuel bills go up. Not green.

The "A" Dimension is the reason why you sometimes see a boiler installed in a pit. The Dead Men didn't dig those holes because they had shovels and time on their hands. They dug them to create the 28" "A" Dimension. If you're replacing that old boiler, keep this in mind. Find that lowest, horizontal, steam-carrying pipe and measure down from it to the center of the gauge glass. If it's a gravity-return system, that distance must be at least 28 inches. If it's more than 28 inches, that's fine. If it's less than 28 inches, you're probably going to have water hammer.

So think twice before you decide to install the new boiler outside the pit. If you abandon the pit, you'll need a boiler-feed pump, and end-of-main steam traps. Both cost money.

Oh, and I mentioned this earlier but it's worth mentioning again. Insulate those return pipes. The cooler the condensate gets, the more carbon dioxide it will absorb, and that creates carbonic acid, which eats steel. It pays to insulate.

Is zoning steam green?

To review: We know that a gravity-return steam system uses a combination of two forces to put the returning condensate back into the boiler. First, there's the steam pressure that's left at the end of the mains. This pressure will never be as great as the pressure that's inside the boiler because, as steam travels through the pipes, it loses some of its energy to friction. What you wind up with at the end of the main depends on the size of the pipes and the boiler's load. The Dead Men figured all of this out years before you were born.

The other force that works to put the returning condensate back into the boiler is gravity. Gravity combines with the leftover steam pressure at the end of the main to create a force that's greater than the pressure inside the boiler. The Dead Men allowed enough vertical space between the end of the lowest steam main and the boiler's water line to give the returning condensate a place to stack up. We call that the "A" Dimension. Combine the two and gravity kicks in. The water slides back into the boiler and all is well with the world.

But now, you're tempted to add motorized zone valves to the steam mains so that you can save some fuel dollars. You figure it makes sense to heat only the parts of the building that you're using.

My experience with motorized steam-main valves is that they work beautifully – as long as they never close. But as soon as one valve closes, that leftover steam pressure that you were depending on to put the condensate back into the boiler disappears from that zone. All you have going for you is gravity at that point, and that's not enough to put the returning condensate back into the boiler. So the condensate backs up into the steam main and lies there, waiting for that motorized valve to reopen. And when it does, the steam comes raging through, meets the water in that horizontal return main, and sends it rocketing toward the end of the main where it hits with enough force to get your attention (and probably to break the pipes).

Perhaps you've heard this noise?

But this isn't the only problem. Keep in mind you have a burner on that boiler that's sized to provide steam for the entire building. When any motorized zone valve shuts, the firing rate doesn't change, does it? Suddenly, you have more steam volume than the pipes can handle. Too much steam volume means the steam will move at a higher velocity, and as the velocity increases, the steam finds it easier to suck water out of the boiler. That, of course, leads to water hammer in the zones that are calling for heat. Go back and look again at those photos of the glass-piped boiler. Watch what happens when Mike opens that ball valve. All the water in the boiler winds up in the parking lot.

But wait, there's more. As the water goes flying into the pipes, the boiler begins to drop into a low-water condition. The burner shuts off, and before the condensate can return (from the system, not the parking lot), the automatic water feeder kicks in and adds water to a system that doesn't really need more water. Then, when the condensate finally does return from the system, the boiler floods.

You'll probably blame the automatic water feeder for this. The feeder is innocent - but it's there, and it's convenient.

Next? Someone decides to install a check valve in the wet return. That person figures that a check valve will keep the water from backing out of the boiler. On the day that the check valve goes in, the water stops backing out of the boiler and it seems as if the problem is solved, but it's really just beginning. Now the condensate can't get back into the boiler because there's not enough pressure to open the check valve. The water hammer continues its psychotic banging whenever any zone valve opens.

Hmm. What to do? What to do?

You install a boiler-feed pump, which costs money and needs maintenance. And by installing the pump, you opened the system's returns to atmosphere because the pump has a big condensate receiver that's vented to atmosphere. Steam now spews into the boiler room from the vent.

Yikes!

You decide to install one huge master trap at the inlet to the boiler-feed pump to keep the steam from venting. This has never once worked in the history of central heating because a master trap allows steam to work its way into the formerly wet returns. Now, you have more water hammer than you had before. Worse yet, whenever all the zone valves close on a boiler that's filled with steam, a vacuum forms inside the boiler. This causes the water that's in your new boiler-feed pump to flow into the boiler and flood it.

Had enough yet?

If you want to zone steam, it's best to do so at the radiators with thermostatic radiator valves. They work.

If you've already installed the big motorized valves on the steam mains, apologize to Mother Nature, and then try this: Run a half-inch line from one side of the zone valve to the other. Make sure you run it over the *top* of the valve. This line will bleed some steam pressure into the mains, and it just might get you off the hook. But then again, it might not.

It's best to pass on the motorized valves.

Get out of the boiler room

There's a lot of piping in a steam system, and it all works together. If you want to save money, take a walk around the building. Think like air and think like steam and condensate as you go. If you were air, could you get out? If you were the steam, which way would you go? If you were condensate, would you be able to drain back to the boiler by gravity? And if not, is there a pump to help you get back?

Take your time and wander around. It pays, and it saves green.

CHAPTER 6

GREENING THE VENTING

Why vents die

Every steam system is an open system. Water expands 1,700 times when it changes to steam, and that shoves the air out. Steam condenses inside the pipes and radiators and shrinks to 1,700th its volume, and that brings the air back. Steam systems breathe – just as we do.

You already know that where there is air, steam will not go. Steam and air have different densities (steam is lighter than air). Steam acts like a plunger, shoving the air ahead of itself. When it condenses, the steam leaves the insides of the pipes all wet. The air enters and reacts with the wet steel and iron, forming rust.

Now, the steam moves so quickly through the pipes that it carries the rust right to the air vents. The vents clog and call in sick. They're not going to work today. When this happens, the other vents have to pick up the slack and work even harder than they usually do. They're *really* blowing air now, and you can hear that air moving at high velocity. Some people think that's a good thing. They hear the air and say, "Yep, she's doing a great job!" But when you're able to hear the air moving from the vent, what you're listening to is the sound of high-velocity air that's shoving even more rust and debris from the pipes into the vents. You shouldn't hear vents venting. It's *not* a good sound.

When the main vents clog (and these handle most of the air in the system), people often replace them with plugs. Plugs don't vent. They plug. Hence their name.

An experiment

This will help you appreciate the importance of main vents. I took a regular plastic bottle and hung a balloon inside of it. So now we ask and answer these important questions:

Q: What's in the bottle?

A: A balloon! (Oh, and air.)

Q: What's inside the balloon?

A: Air!

Q: Can I blow up that balloon?

A: Um, maybe a *little* bit.

Q: Why just a *little* bit?

A: Because there's air on both the inside and the outside of the balloon. If the balloon wasn't there, I really couldn't blow that much more air into the bottle, could I? Just what my lungs could compress, and that's not much.

Get it? If the air doesn't have a way out of the bottle, you can't add much more air to the bottle because you don't have that much pressure available in your lungs. Steam systems operate at low pressure, just like your lungs. If the air doesn't have a way out of the pipes and radiators, you won't be able to add steam to the pipes and radiators. Where there is air, steam will not go.

So let's give the air a way out.

We'll poke a hole in the bottom of the bottle. Now air can vent. Let's try blowing up the bottle again.

Hole in bottom of bottle

Happy to provide!

I'm full of hot air so this part is easy. The air goes out and the balloon expands. Now think of the balloon as the steam and you'll understand the importance of those main vents in a steam system. When those main vents clog, or when somebody replaces them with plugs, the fuel bills go up because some knucklehead is bound to raise the pressure.

Meet the knucklehead

This is the guy who shows up with the clockwise screwdriver. The steam isn't moving down the mains and he thinks he knows why. He figures it's because the mains are clogged. He's right about that, but it's *air* that's doing the clogging. If he could, the knucklehead would pour drain cleaner into the radiators and let it flow down into the pipes, but this would take too much effort.

So he uses his clockwise screwdriver to raise the steam pressure. He's going to blow out the clog using steam pressure. He does this because he's a knucklehead.

And it seems to work because of the First Law of Steam Heating (*When you do something stupid, you will always get a reward, which leads you to do things of even greater stupidity*). The knucklehead goes back to the main where he suspected there was a clog and he feels the pipe. Steam has moved a bit further down the line. Eureka! What the knucklehead doesn't understand, though, is that the steam is only compressing the trapped air. The air is getting aggravated at being pushed around and it's pushing back. That's what's causing the pressuretrol to shut off the burner, even at the higher pressure.

But the knucklehead sees only the results, however meager, and he finds joy in that. The steam moved! So he goes back to the boiler and raises the pressure even higher.

Eureka!

And he does this again and again until the burner runs 24 hours a day, while the building stays cold, and the fuel bills soar.

Again, you can move with a pinhole what you can't move with a ton of pressure. Think like air and walk along the pipes. Ask that critical question: If I were air, could I get out?

Don't be a knucklehead.

Where do vents belong?

Let's start with one-pipe-steam radiator vents. The vent belongs on the side opposite the supply valve, and in a tapping that's partway down from the top of the radiator. Like this:

Steam is lighter than air, so when it enters a one-pipe radiator it heads straight for the top. If you have the air vent on the side opposite the supply valve, but right at the top of the radiator, the steam will shut the vent before most of the air can get out of the radiator. And as you know, if air can't get out, steam can't get in.

It's best to place the vent partway down the radiator. You should see a plug there for the vent (if the vent's not already in place), or at least a boss where you can tap for the vent. Place it here and the radiator will heat more fully.

For very large, one-pipe radiators, the Dead Men used two vents. Like the radiator pictured below.

On start-up, both vents vented. When steam reached the upper one, the venting rate would slow, and eventually stop when steam reached the lower vent. They did this because a

Where the air vent belongs

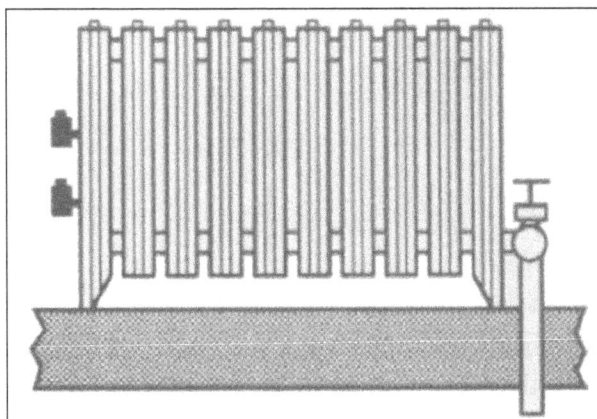

very large, one-pipe radiator can produce a lot of condensate on start-up. They didn't want that condensate fighting the incoming steam as it tried to flow back to the boiler, and this venting trick helped with that. It slowed the heating process a bit, and it works.

If you want to use thermostatic radiator valves, go back and read that section of the book where we talked about these (it's in the chapter on Greening the Load).

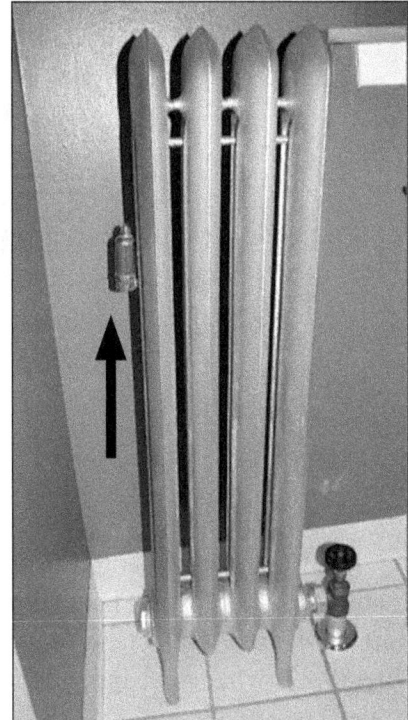

Radiator with two vents

If you're working with a very old, one-pipe system, watch out for vacuum-type air vents. They'll have a tiny check valve in the vent hole, and they'll probably say "vacuum" on the vent body. Look for that. Vacuum vents depend on a coal fire to work properly. If you're burning oil or gas, vacuum vents will cause balance problems in the system. If you see them, replace them with standard one-pipe-steam vents.

You're also going to need a good main vent (or maybe more than one) near the end of each horizontal steam main, just before it drops vertically to return to the boiler. Here's how to pipe them.

Properly piped main vent

Ideally, the vent should be about 15 inches back from the end of the main, and about 10 inches up on a nipple. This will protect it from any water hammer that may happen when steam and condensate hit the end of the line. If you install (or see) a main vent right at the end of the line, this is what can happen to it.

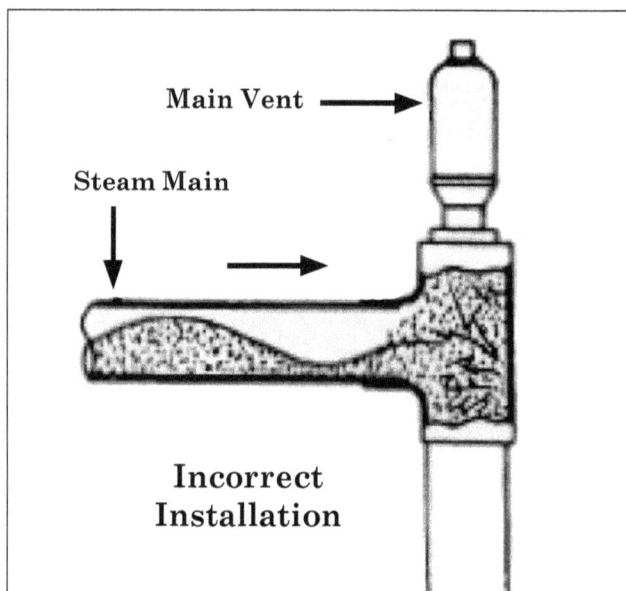

One shot of water hammer is all it takes to ruin a good main vent. And once the vent is broken, it's not going to vent air very well. And if the air can't get out, the steam can't get in. Bad main vents cost fuel dollars.

And I know that you can't always get a main vent in that exact perfect spot, but use that photo as a guide and get it as close as you can, working within the space you have on the job.

Master venting

When steam heating was new, coal was the fuel of choice. Coal burns for hours and hours, making a slight pressure all the while. With this near-constant pressure, the steam was always pushing at the air, nudging it from the main vents and the radiator vents. The only time the pressure shut off was when the coal fire died down.

Because of this, the Dead Men ran their steam mains around the perimeter of the basement, feeding the radiators on all the floors along the way, and draining the condensate from the end of that perimeter main. The main just bumped off the corners of the basement and all was well with the world.

Then the 1930s arrived. Coal was expensive because they had to mine it and ship it, and it was also a heavy and dirty fuel that had to be handled. Homeowners and commercial-building owners had to tote the ash and deal with the clinkers and it was a huge amount of work. Besides, we were in the Great Depression and that made things even tougher.

And then oil arrived and it was not only cheaper than coal, it was automatic. You didn't need a bin for the oil. You didn't have to shovel it. You didn't need a damper regulator to control the fire. And compared to coal, it was immaculate. That really appealed to housewives, who, for the most part, were the ones responsible for building the fire each day. And the best part was an oil fire worked with a thermostat! It was totally automatic. People lined up to make the switch to oil (or in some cases, natural gas) and coal became a relic.

And that's when the problems began. Most steam systems had those perimeter mains. The Dead Men had piped those jobs for the uniqueness of a coal fire that never really shut off. With an automatic fuel, the venting was all wrong. The thermostat was somewhere in the middle of the building, and it shut off the burner before most of the air was out of the main. The building would be warm on one side, and cold on the other. It was a big problem, and the oil dealers solved it by piping their new jobs with multiple horizontal mains, balancing one against the other. This was when we began to see headers on steam boilers with multiple take-offs for the mains. Watch for this, and for those perimeter mains, when you're roaming basements. It will give you a good sense of the system's age, and whether or not the Dead Men originally piped it for coal or for an automatic fuel such as oil or gas.

On the older jobs (those that were piped for coal) the oil men solved the balance problem by using very large main vents. They knew that if they could vent the main very quickly, the steam would favor that route over the risers. They would be able to fill the horizontal main like a trough, because steam will always follow the path of least resistance. By using those large main vents, the Dead Men coaxed the steam into arriving at all the risers that led to all the radiators at about the same time.

I hope you're beginning to see how important those main vents are – and how they have to be just the right size. And this brings me, once again, to Frank.

Frank Gerety

I started telling you about Frank Gerety back in the Greening the Pressure chapter. As he was doing his work in New York City with the Department of Housing Preservation and Development during the 1980s, he was writing mainly about the problems caused by wet steam. He was coming up with new ways to pipe steam boilers to dry the steam, and his focus was mainly on apartment buildings of all sizes.

Frank also realized that trapped air was a big problem in buildings of this type. He knew that steam wants to move at 60 mph, and the only thing standing in its way is air. So he began to propose a technique he called Master Venting. I watched his work as it evolved and it amazed me. In the Master Vented buildings, the steam traveled faster than I could run. These systems went into balance with just ounces of pressure at the boiler, and the burners didn't short-cycle.

This was when we all began to see the low-hanging fruit.

The challenge, though, was that the air-vent manufacturers were telling people to vent by geography. They said that radiators that are closest to the boiler should have slower air vents than radiators that are furthest from the boiler. The more Frank thought about this, the less sense it made to him.

Here, consider this: Big rooms have large radiators because they have a large need for heat. Small rooms have small radiators because they have less need for heat. On the coldest days of winter, we want both the large and the small radiators to heat from one end to the other, and we want that to happen at the same time. When it does, the system is in balance. When it doesn't, people either open windows to let out the excess heat, or they call to complain. Not good either way.

It really doesn't matter where the radiator is within the building. What matters is system balance. All the radiators should heat fully on the coldest days of the year. It's as simple as that. It's not about radiator location; it's about the amount of air to be vented from each section of the system.

It seemed to us that the vent valve manufacturers had it all wrong.

The Dead Men's Steam School

In 1989, Dick Koral, who was the Director of New York City's Apartment House Institute, gave me 16 file drawers of data, which his old boss, Clifford Strock, had collected over the years. I never got to meet Mr. Strock. He was the publisher of *Air Conditioning, Heating and Ventilating* magazine. He was also a collector. Dick was his editor. The files that Dick gave me contained hundreds of original magazine articles and product pamphlets from the 1910s up through the 1960s. It was a treasure trove of information, and, in a way, a time machine. I got to read about problems the Dead Men were having with steam systems in their own words.

It was wonderful. One of those articles explained the problems the oil men were having with those jobs that had been piped to run on coal. That's how I knew.

I took that knowledge, along with what I had learned from watching Frank Gerety do his remarkable work in the apartment buildings of New York City, and I put together a seminar called the Dead Men's Steam School.

An engineer friend had inspired this one. For years, I had done seminars on how to troubleshoot steam systems, and how to get the most out of them, but he wanted to know the *whys* of steam systems. "I wish you would do a trade-school sort of seminar," he said. "Make believe it's 1920 and your students are going to size, design, and install a steam system from scratch. What do we need to know? How and *why* do we do each thing? Take it step by step."

And that's what I put together. We began the seminar with an imaginary empty apartment building. We figured the heat loss, the radiator sizes, the piping, and the venting.

Back then, we were all working with the venting rates that the air-vent manufacturers had provided in their literature. That information had been around for many years and it seemed fine to me.

I knew from Frank Gerety, and from those old magazine articles that the key to Master Venting was to treat the steam mains, the risers, and the radiators as separate venting challenges. You couldn't take them all together if you wanted the system to balance. Air was going to follow the path of least resistance as it left the system, and steam was going to follow the air, so we had to create those paths of least resistance.

It works like this:

1. You measure the air in the mains and size the main vents to handle *only* that air.

2. If there are multiple steam mains (there usually are), you measure the air in each of them, and then you size the individual main vents so that one balances the other. In other words, if one main contains 100 cubic feet of air, and a second, shorter, main, contains just 50 cubic feet of air, you'll need twice the venting capacity on the first main so that both mains vent at the same rate. Make sense?

3. You do the same with the risers that lead from the steam mains to the radiators. You're going to vent the tops of those risers. The steam will reach the base of all the risers at about the same time because you did such a fine job of venting the mains. Balance one riser against the others so that the steam climbs each riser at the same rate. In other words, do for the risers what you did for the mains.

4. Next, measure the air that's in a runout to a radiator on a riser, and add to this the air that's inside that radiator. Size your radiator vent based on that amount of air.

5. Do this for all the radiators and balance one against the other. Big radiators should vent faster than small radiators, regardless of where they are in the building because big radiators contain more air.

I taught this class for several years and then had it videotaped. We sold the video through the HeatingHelp.com Web site (it's no longer available). It went all over the country and lots of professionals started to balance steam systems, and many had results such as those Frank Gerety enjoyed.

Enter Gerry Gill and Steve Pajek

Gerry and Steve operate G.W. Gill Plumbing & Heating, which is located at 80 W. Grace Street, Bedford, OH 44146. You'll find them on the Web at www.gwgillplumbingandheating. com, and that's worth the keystokes. You can also call them at 440-439-4417.

I'm going to step back for a moment and let Gerry tell you their story in his own words.

"A while back, Steve Pajek and I had a reason to question how much air a radiator trap could pass, because, until steam hits a radiator trap, that trap is an air vent, or at least it's a venting means.

"This led us to procure a quantity of Dwyer Company flow meters and start testing devices to determine their venting capacities. Of course, we couldn't resist testing actual air vents also. What came out of these tests is the material presented in the venting-capacity chart.

"We presented our findings to Dan Holohan, who asked us to put together our method of using this information so that it would be presentable as more than just numbers. As he put it, 'The numbers have to be used for something or they are just numbers.'

"So here is our methodology to balancing steam systems, but first, we would like to say that very little here is an original idea. Our methodology is compiled from those who share their insights on HeatingHelp.com's community Web page known as The Wall. To Dan Holohan and all the Wallies, we say thank you."

Gerry and Steve had done a remarkable bit of research. They questioned the venting rates that I had been using for years (the ones from the Dead Men's Steam School, which the manufacturers had provided). It just wasn't working as well as they thought it should in the field. So they tested the vents and the traps and I have their permission to share with you what they found.

And as a testament to how good these two guys are, when they put together their findings in an e-booklet called, ***Balancing Steam Systems Using a Vent-Capacity Chart***, they gave it to me to sell on the HeatingHelp.com Web site. At their request, *all* of the proceeds from that e-booklet go to a literacy program at New York Cares, which is a volunteer organization in New York City that helps those less fortunate. A portion of the profits of this book also goes to that same literacy program. But let me now turn the floor back to Gerry.

"Steve and I are firm believers in venting the steam main pipes as quickly as possible, and this includes the vertical main to perhaps a third floor. We don't feel that steam knows horizontal from vertical. And if you desire to have adequate heat on the third floor as quickly as on the first floor, then the vertical riser needs a main vent at its top. So, our first step is to vent the mains, and we learned this lesson from Noel Murdough. He may not even be aware that he taught us this, as its been so long that we don't remember if he gave it to us directly, or just implied it. That's the power of The Wall. In any event, we've come to refer to this as the Noel Method. Here it is."

Let me jump in here for a moment to tell you a bit about Noel. He's done many things in his career and he's one of the best steam men I've ever met. He's always willing to share what he knows and you can get to know him a bit better by visiting his blog on the Web at http://steamheating.googlepages.com/home. Back to Gerry:

"Take the vent off the end of the main and fire the boiler. Once the header pipe gets hot, time how long it takes for steam to get to the open pipe where you removed the vent. Let's say it takes three minutes, for example. If it takes three minutes to get steam from the header to the open pipe, then you would need to install as many main air vents as it takes, to get steam to the same point (the open pipe), in as close to three minutes as possible, but with the air vents installed. For example, if it takes three minutes with an open pipe, and it takes six minutes with a Gorton #2 air vent in place, you would add another Gorton #2 and time it again. If it now takes 3-½ minutes to get steam to the end of the main with two Gorton #2 air vents in place, that's only 30 seconds more than what you'd get from an open pipe.

"At this point, you have to make a decision as to the cost of another main vent, vs. a 30-second increase in speed. Why would you get only 30 more seconds speed if adding another vent, when adding a second vent gave you 2-½ minutes of speed increase? Because you will not be able to vent any faster than the open pipe did, regardless of how many air vents you use at this point. You would be venting the main as quickly as possible with the existing tapping. The only way to increase the speed beyond this point would be to increase the diameter of the tapping, say, from ¾" to 1", or to add another tapping and start all over again. You would then do the vertical main in the same manner."

Okay, this is a good time for me to show you a couple of charts. First, there's this one, which comes from the Dead Men's Steam School. It's very easy to use. All you need is a tape measure and a pipe-measuring tool.

Pipe Size	Cubic feet of air per linear foot of pipe
1-1/4"	.010
1-1/2"	.014
2"	.023
2-1/2"	.03
3"	.053
3-1/2"	.07
4"	.09
5"	.14
6"	.2
8"	.36

If you're not sure of the size of the main (or if it's covered with insulation), consider getting this tool called the Pocket Rocket.

It's a caliper that reads the diameter of pipes in inches. It's not going to be exact to the NPT size, but it's very close, and you'll know what you have. If there's insulation on the pipe, you can slit it with a sharp knife and slide the Pocket Rocket in. I've used one of these for years. It's a terrific tool and you can get one in the **Shop** at HeatingHelp.com

And now we're ready for the chart that Gerry and Steve came up with after testing the most common main vents used today, and lots of obsolete ones that may still be on the job. This is all original research, and *very* good information to have.

Main Vents	CFM @ 1 ounce	CFM @ 2 ounce	CFM @ 3 ounce
Open 1/8" steel pipe	1.200	2.000	2.500
Open ½" steel pipe	2.600	3.400	4.800
Open ¾" steel pipe	*Note #1	*Note #1	9.500
Open 1" steel pipe	*Note #1	*Note #1	11.000
American Radiator Co. Ideal Vac-Vent #822 (obsolete)	0.100	0.158	0.200
Barnes & Jones QV1 (obsolete)	0.583	1.580	2.000
Barnes & Jones Ventrite #75 (obsolete)	0.116	0.175	0.225
Detroit Controls #841 Quick Vent (obsolete)	0.333	0.550	0.700
Dole #5 Quick Vent	0.066	0.116	0.150
Gorton #1	0.330	0.540	0.700
Gorton #2	1.100	1.750	2.200
Hoffman #4	0.060	0.110	0.130
Hoffman #4A	0.133	0.216	0.266
Hoffman #6 Vacuum Valve (obsolete)	0.033	0.083	0.108
Hoffman #10 Vapor Vent (obsolete)	2.033	2.500	3.700
Hoffman #11 Vapor Vacuum Vent (obsolete)	1.000	2.410	3.330
Hoffman #15 Vacuum Vent for Differential Loop (obsolete)	1.283	2.750	3.416
Hoffman #16 Vacuum Valve (obsolete)	0.066	0.116	0.150
Hoffman #16A Vacuum Valve (obsolete)	0.100	0.183	0.241
Hoffman #75	0.500	0.750	0.960
Hoffman #76 Vacuum Valve	0.383	0.666	0.833
Ideal Heating Co. "Wafer Style" (obsolete)	0.090	0.130	0.166
Maid-O-Mist #1 Straight Pattern	0.333	0.590	0.766
Marsh #55	0.010	0.021	0.033
Mouat Cast Iron "Bullet" Style (obsolete)	0.530	0.910	1.110
Mouat Tri-Tube (obsolete)	1.160	1.800	2.000

continued from previous page...

Main Vents	CFM @ 1 ounce	CFM @ 2 ounce	CFM @ 3 ounce
Squires Co. Mouat Series Cast Iron Bullet Style (obsolete)	0.530	0.910	1.110
Trane Cast-Iron Bullet Style (obsolete)	0.780	1.330	1.700
Vent Rite #35	0.110	0.200	0.250
Vent Rite #75 – Barnes & Jones issue (obsolete)	0.116	0.175	0.225
Vent Rite #77	0.190	0.310	0.400
Watts SVS-3	0.070	0.116	0.150
Warco #3 (obsolete)	0.050	0.108	0.150
Warco #87 (obsolete)	0.108	0.175	0.233
Warren Webster #40-15 Vent Trap (obsolete)	1.166	1.866	2.366

*Note 1 Due to the size of the flowmeter and pipe, at least 3 ounces of pressure was required to produce a reading

Here's Gerry again:

"At this point, you may be wondering what do the capacity numbers have to do with all of this. Well, it gives you an idea of where to start. Consider that ¾" pipe tapping, for example. If you want to achieve the same speed as you would get from that open pipe, then you have to be able to vent the same amount of cubic feet, right? So, if a wide-open, ¾" pipe can vent 9.5 cubic feet per minute (at 3 ounces of pressure from the boiler), then your starting point would be either to use four Gorton #2 air vents, or 10 Hoffman #75 air vents, or 23 Ventrite #77 air vents. It's your choice, and your dollars.

"Now here's how the manifolding of main vents and the chart of capacities work together. Picture a ½" pipe tapping with a 90-degree elbow and two Gorton #2 vents on the horizontal manifold. The combined venting of the two Gorton #2 vents (at 2 ounces of pressure from the boiler) would be 3.4 cfm (1.7 each), instead of the 1.75 cfm each is capable of venting separately. Why? Because the ½" pipe can only deliver 3.4 cfm. Divide that by two and you get 1.7.

"The same setup, but with two Gorton #2 air vents on a ¾" pipe, will vent at 3.5 cfm at 2 ounces of pressure from the boiler. That's more than a wide-open ½" pipe will give you, and it's equal to the capacity of both vents at that pressure. See how it works? With the ½" pipe, we were limited in venting capacity by the pipe's ability to deliver cfm. With the ¾" pipe, we have plenty of pipe capacity, and the only thing limiting us now is the venting rate

of the vents themselves. Or to put it another way, using more than two Gorton #2 air vents on a ½" pipe would be a waste of time and money.

"A ¾" pipe at 3 ounces pressure from the boiler will pass 9.5 cfm. A Gorton #2 air vent at 3 ounces of pressure will pass 2.2 cfm. So 9.5 cfm, divided by 2.2 cfm gives us a required total of 4.31 Gorton #2 vents. Now if four Gorton #2 air vents deliver 8.8 cfm at 3 ounces of pressure, is the fifth Gorton #2 vent worth the money to gain just .7 cfm? We don't think so, but a Gorton #1 as your fifth air vent would max out the cfm potential, and that would be a good choice. Simple, isn't it?"

Get it? You vent the horizontal mains so that the steam reaches the end in a certain amount of time, which you can determine by measuring the air contained in the mains and sizing your main vent (or vents), and the tapping that serves those main vents, accordingly. The steam will reach the base of all the risers at about the same time. You then measure the air in the risers and balance one against the other by sizing the riser vents, which you'll pipe at or near the top of the riser. If the riser ends at a radiator, you may be able to replace the radiator's supply valve with a tee. Put your riser vent in the run of that tee, and repipe the radiator off the bull of the tee. You'll probably only have to move the radiator a few inches. This was one of Frank Gerety's ideas and here's his sketch.

Okay, now you have steam at the top of all the risers, which means it's ready to enter the short pipe that connects each radiator to those risers. Measure the air that's in that pipe and add to this the air that's in the radiator. This will give you what you need to choose the proper radiator vent. Balance the big radiators against the smaller radiators by venting from both at different rates. Big radiators contain more air than small radiators so big radiators should have air vents of greater capacity than those you'll use on the smaller radiators.

You'll need this chart again to know how much air you're dealing with in the pipe leading from the riser.

Pipe Size	Cubic feet of air per linear foot of pipe
1-1/4"	.010
1-1/2"	.014
2"	.023
2-1/2"	.03
3"	.053
3-1/2"	.07
4"	.09
5"	.14
6"	.2
8"	.36

And you'll need this chart to know how much air is in the most-common radiators. You'll first have to know the radiator's EDR rating, of course, and you can look back to the chapter on Greening the Load for a review of how that works.

Radiator type	Cubic feet of air per square foot EDR
Cast-iron flue (antique radiator)	.029
Cast-iron wall- or ceiling-mounted	.028
Cast-iron column (circa-1900)	.025
Cast-iron tube (circa-1930)	.013
Cast-iron thin tube (as made today)	.009
Cast-iron radiant radiator (5" deep)	.008
Cast-iron baseboard (10" high)	.07
Cast-iron convector (in cabinet)	.0003
Unit heater	.0002

Here's the chart that Gerry and Steve put together for sizing your radiator vents.

Radiator Vents	CFM @ 1 ounce	CFM @ 2 ounces	CFM @ 3 ounces
Automatic #1 (obsolete)	0.045	0.222	0.100
American Radiator Co. Detroit Controls #500 In-Air-Rid (obsolete)	0.036	0.061	0.080
ARCO Detroit Controls Multiport Setting 1	**Note 3	**Note 3	**Note 3
ARCO Detroit Controls Multiport Setting 2	**Note 3	**Note 3	**Note 3
ARCO Detroit Controls Multiport Setting 3	**Note 3	**Note 3	0.011
ARCO Detroit Controls Multiport Setting 4	0.021	0.028	0.038
ARCO Detroit Controls Multiport Setting 5	0.045	0.075	0.100
ARCO Detroit Controls Multiport Setting 6	0.050	0.108	0.141
ARCO Detroit Controls Multiport Setting 7	0.066	0.116	0.158
ARCO Detroit Controls Multiport Setting 8	0.083	0.141	0.191
ARCO Detroit Controls Multiport Setting 9	0.100	0.158	0.208
ARCO Detroit Controls Multiport Setting 10	0.116	0.183	0.241
Detroit Controls #5000 Airid Variport @ Low setting (obsolete)	NONE	0.011	0.016
Detroit Controls #5000 Airid Variport @ Medium setting (obsolete)	0.036	0.051	0.061
Detroit Controls #5000 Airid Variport @ High setting (obsolete)	0.066	0.125	0.166
Dole 1933	0.033	0.055	0.070
Dole 1A, Setting 1	0.045	0.071	0.125
Dole 1A, Setting 2	0.068	0.116	0.150
Dole 1A, Setting 3	0.083	0.125	0.166
Dole 1A, Setting 4	0.091	0.150	0.191
Dole 1A, Setting 5	0.125	0.208	0.266
Dole 1A, Setting 6	0.133	0.216	0.283

continued from previous page...

Radiator Vents	CFM @ 1 ounce	CFM @ 2 ounces	CFM @ 3 ounces
Dole 1A, Setting 1 – Modern version with plastic tongue *Note 4	0.061	0.111	0.148
Dole 1A, Setting 2 – Modern version with plastic tongue *Note 4	0.084	0.141	0.186
Dole 1A, Setting 3 – Modern version with plastic tongue *Note 4	0.108	0.183	0.228
Dole 1A, Setting 4 – Modern version with plastic tongue *Note 4	0.116	0.201	0.254
Dole 1A, Setting 5 – Modern version with plastic tongue *Note 4	0.129	0.210	0.266
Dole 1A, Setting 6 – Modern version with plastic tongue *Note 4	0.133	0.220	0.274
Dole 1A, Setting 7 – Modern version with plastic tongue *Note 4	0.143	0.233	0.295
Dole 1A, Setting 8 – Modern version with plastic tongue *Note 4	0.147	0.245	0.283
Dole 1A, Setting 9 – Modern version with plastic tongue *Note 4	0.152	0.268	0.289
Dole 1A, Setting 10 – Modern version with plastic tongue *Note 4	0.152	0.269	0.289
Dole 1A, Cap removed – Modern version with plastic tongue *Note 4	0.212	0.339	0.431
Flair #51 (obsolete)	0.116	0.191	0.250
Gorton 4	0.025	0.040	0.055
Gorton 5	0.080	0.130	0.160
Gorton 6	0.150	0.235	0.300
Gorton C	0.270	0.450	0.570
Gorton D	0.330	0.540	0.700
Heat Timer Varivalve Minimum Setting, Angle Pattern	0.065	0.158	0.200

continued from previous page...

Radiator Vents	CFM @ 1 ounce	CFM @ 2 ounces	CFM @ 3 ounces
Heat Timer Varivalve 50% Setting, Angle Pattern	0.340	0.580	0.766
Heat Timer Varivalve Maximum Setting, Angle Pattern	0.516	0.850	1.130
Heat Timer Varivalve Minimum, Setting Straight Pattern	0.060	0.083	0.330
Heat Timer Varivalve 50%, Setting Straight Pattern	0.530	0.800	1.030
Heat Timer Varivalve Maximum Setting Straight Pattern	0.660	1.080	1.360
Hoffman #1 (obsolete)	0.016	0.075	0.100
Hoffman 1A, Setting 1	0.020	0.026	0.033
Hoffman 1A, Setting 2	0.026	0.043	0.056
Hoffman 1A, Setting 3	0.100	0.158	0.200
Hoffman 1A, Setting 4	0.108	0.160	0.210
Hoffman 1A, Setting 5	0.140	0.220	0.290
Hoffman 1A, Setting 6	0.145	0.225	0.300
Hoffman #2 Vacuum Vent Non-Adjustable (obsolete)	0.016	0.066	0.100
Hoffman #2 Vacuum Vent, Setting 1 (obsolete)	0.001	0.021	0.025
Hoffman #2 Vacuum Vent, Setting 2 (obsolete)	0.006	0.024	0.028
Hoffman #2 Vacuum Vent, Setting 3 (obsolete)	0.008	0.025	0.033
Hoffman #2 Vacuum Vent, Setting 4 (obsolete)	0.010	0.026	0.036
Hoffman #2 Vacuum Vent, Setting 5 (obsolete)	0.013	0.029	0.040
Hoffman #2 Vacuum Vent, Setting 6 (obsolete)	0.016	0.031	0.045
Hoffman #2A Vacuum Vent, Setting 1 (obsolete)	0.001	0.023	0.025
Hoffman #2A Vacuum Vent, Setting 2 (obsolete)	0.018	0.030	0.043

continued from previous page...

Radiator Vents	CFM @ 1 ounce	CFM @ 2 ounces	CFM @ 3 ounces
Hoffman #2A Vacuum Vent, Setting 3 (obsolete)	0.021	0.038	0.053
Hoffman #2A Vacuum Vent, Setting 4 (obsolete)	0.028	0.058	0.068
Hoffman #2A Vacuum Vent, Setting 5 (obsolete)	0.036	0.063	0.088
Hoffman #2A Vacuum Vent, Setting 6 (obsolete)	0.041	0.083	0.116
Hoffman #3 (Paul vent)	0.175	0.283	0.416
Hoffman #40	0.042	0.067	0.087
Hoffman #41	0.058	0.100	0.125
Hoffman #70 Airport Non-Adjustable (obsolete)	0.050	0.108	0.141
Hoffman #70 Airport, Setting 1 (obsolete)	0.045	0.078	0.125
Hoffman #70 Airport, Setting 2 (obsolete)	0.050	0.087	0.150
Hoffman #70 Airport, Setting 3 (obsolete)	0.053	0.116	0.166
Hoffman #70 Airport, Setting 4 (obsolete)	0.056	0.141	0.183
Hoffman #70 Airport, Setting 5 (obsolete)	0.060	0.150	0.191
Hoffman #70 Airport, Setting 6 (obsolete)	0.066	0.150	0.191
Hoffman #74 Unit Heater Vent	0.033	0.083	0.108
Hoffman #500	0.066	0.112	0.142
Homart #775.8812	0.083	0.133	0.175
Marsh Paul Vent, circa-1901 (obsolete)	0.350	0.583	0.750
Marsh Paul Vent #3 (obsolete)	0.183	0.295	0.375
Maid-O-Mist 4	0.028	0.045	0.060
Maid-O-Mist 5	0.100	0.158	0.200
Maid-O-Mist 6	0.150	0.241	0.300
Maid-O-Mist C	0.283	0.450	0.583
Maid-O-Mist D	0.341	0.600	0.783
M.S. Little Co. #L-15 (obsolete)	0.026	0.045	0.055

continued from previous page...

Radiator Vents	CFM @ 1 ounce	CFM @ 2 ounces	CFM @ 3 ounces
National Steam Specialty Co. Paul Vent (obsolete)	0.150	0.226	0.333
U.S. Radiator #3 Triton (obsolete)	0.023	0.033	0.041
Ventrite #1, Setting 1	Off	Off	Off
Ventrite #1, Setting 2	0.033	0.020	0.021
Ventrite #1, Setting 3	0.025	0.036	0.046
Ventrite #1, Setting 4	0.030	0.053	0.066
Ventrite #1, Setting 5	0.045	0.071	0.091
Ventrite #1, Setting 6	0.056	0.091	0.116
Ventrite #1, Setting 7	0.070	0.108	0.133
Ventrite #1, Setting 8	0.083	0.125	0.158
Ventrite #2 IVS (obsolete, Anderson Manufacturing)	0.028	0.053	0.078
Ventrite #11	0.060	0.100	0.116
Watts SV	0.066	0.108	0.133
Watts SVA, Setting 1/8	0.030	0.063	0.075
Watts SVA, Setting 1/4	0.066	0.108	0.133
Watts SVA, Setting ½ ** (see Note 2)	0.066	0.108	0.133
Watts SVA, Setting ¾ ** (see Note 2)	0.066	0.108	0.133
Watts SVA, Setting 7/8 ** (see Note 2)	0.066	0.108	0.133
Watts SVA, Setting Full ** (see Note 2)	0.066	0.108	0.133

** Note 2 – This Watts SVA, made in Taiwan and brand-new out of the box gave these readings. This is not a typographical error.

** Note 3 – The venting rate at this setting was so insignificant as to be imperceptible on the flow meter.

** Note 4 – The anomaly is caused by the fact that there is a screw on top of the lid. This injects a human factor. Depending on how tight one makes the screw, this affects the ability of the device to vent. This data was the average of four vents tested. The Cap-Removed readings are probably the most accurate, as a human factor is not in the equation.

Gerry and Steve go into a lot more detail in their e-booklet, ***Balancing Steam Systems Using a Vent-Capacity Chart***. They take you through several step-by-step examples, with diagrams of the system, and they tell of their experience in the field. You can get a copy of that e-booklet from the **Shop** at www.HeatingHelp.com, and if you choose to buy it, we'll donate 100% of that money to New York Cares.

In that booklet, Gerry and Steve thank John Shea of Grosse Point Farms Michigan for his assistance in this project. John sent them many pieces and parts from old steam systems in and around Detroit so that they could test and rate them.

I thank all three of these great guys.

Balancing two-pipe steam

Gerry Gill has a terrific quote, "Before a radiator trap is a radiator trap, it is an air vent." He is so right, and this is one of the finer points of steam heating that many people miss these days as they repair steam traps on radiators and F&T traps on steam mains or riser drips.

All the air in the pipes and radiators must pass through the traps in a two-pipe system, and that trap's ability to pass air on to a vent or a vented condensate receiver helps determine how quickly steam gets from the boiler to the radiators.

Steam traps are *very* green, as you'll see in the next chapter, but first, consider how much air will move through them. Here's more of Gerry and Steve's work:

Venting Rates of Thermostatic Radiator Traps	CFM @ 1 ounce	CFM @ 2 ounces	CFM @ 3 ounces
Barnes & Jones, ½" #122	1.50	1.800	2.700
Dunham#1, ½" (obsolete)	1.450	2.410	3.000
Dunham #1 with a Barnes & Jones Cage Unit	0.730	1.160	1.500
Dunham-Bush #1E (obsolete)	1.080	1.800	2.250
Dunham-Bush #1E with a Barnes & Jones Cage Unit	0.710	1.130	1.410
Hoffman 17C, ½" with Dura-Stat Capsule	0.530	0.850	1.100
Hoffman 17C, ½" with a Barnes & Jones Cage Unit	1.300	2.116	2.700
Hoffman 17D, ½", Model 1939 (obsolete)	1.450	2.380	3.000
Ideal Heating Company Thermostatic ½" (obsolete)	1.530	2.330	3.080

continued from previous page…

Venting Rates of Thermostatic Radiator Traps	CFM @ 1 ounce	CFM @ 2 ounces	CFM @ 3 ounces
Ideal with a Barnes & Jones Cage Unit	1.250	2.000	2.580
Illinois, ½" #1G	0.860	1.200	1.600
Marsh #1 Reflux (obsolete)	1.130	1.830	2.360
Marsh #1 Reflux with a Barnes & Jones Cage Unit	0.730	1.160	1.500
Marsh 1N, ½"	0.530	0.910	1.160
Mepco, ½" #1E	1.080	1.400	1.900
Mepco, ½" #1R	1.080	1.400	1.900
Milvaco Model #0 (obsolete)	1.080	1.710	2.130
Milvaco Model #0 with a Barnes & Jones Cage Unit	1.000	1.580	1.966
Monash #30, ½"	0.860	1.200	1.600
Monash #34, ½"	1.210	2.000	2.500
Monash #38, ½"	1.450	2.450	3.080
Mouat Model 1924 Thermostatic, ½" (obsolete)	1.530	2.330	3.080
Mouat Model 1924 with a Barnes & Jones Cage Unit	1.250	2.000	2.580
Mouat #35, ½" (obsolete)	1.330	1.660	2.080
Mouat #35, ½" with a Barnes & Jones Cage Unit	1.366	2.330	2.910
Mouat #36, ½" (obsolete)	1.083	1.533	1.830
Mouat #36, ½" with a Barnes & Jones Cage Unit	1.266	2.000	2.500
Mouat Water-Seal Trap, ½" Oval (obsolete)	2.000	2.400	3.400
Mouat Water-Seal Trap, ½" Round (obsolete)	2.000	2.400	3.600
Nicholson #N125	1.530	2.530	3.080
Sarco A125, ½"	0.500	0.780	1.000
Sterling 7-50-A, ½"	1.080	1.750	2.160

continued from previous page…

Venting Rates of Thermostatic Radiator Traps	CFM @ 1 ounce	CFM @ 2 ounces	CFM @ 3 ounces
Trane B1 with Original Bellows (obsolete)	0.500	0.750	1.000
Trane B1 with Modern Trane Element	0.780	1.250	1.580
Trane B1 with a Barnes & Jones Cage Unit	0.750	1.200	1.500
Trane ½" – No Number (obsolete)	1.580	2.660	3.330
Warren Webster 02H, ½" with a Barnes & Jones Cage Unit	0.783	1.333	1.750
Warren Webster 02H, ½" with a Tunstall Capsule	0.616	0.966	1.210
Warren Webster 512H, ½" with 512A seat (5-/16") (obsolete)	1.080	1.750	2.250
Warren Webster, 512H, ½" with 512S seat (1/4"), (obsolete)	0.783	1.280	1.660
Warren Webster 512H, ½" with a Barnes & Jones Cage Unit	1.360	2.250	2.830
Warren Webster 712HB, ½" with H57 seat (.368") (obsolete)	0.750	1.233	1.433
Warren Webster 712HB, ½" with 522S seat (.332") (obsolete)	0.500	0.800	1.250
Warren Webster 712HB, ½" with a Barnes & Jones Cage Unit	1.550	2.550	3.216
Barnes & Jones ¾" #134A	1.360	2.280	2.880
Hoffman 8C, ¾"	1.000	1.580	2.000
Mepco 2E, ¾"	1.530	2.550	3.160
Monash #48, ¾"	2.160	3.500	5.000
Sterling #753A, ¾"	1.250	2.000	2.580

Venting Rates of Float & Thermostatic Traps	CFM @ 1 Ounce	CFM @ 2 Ounces	CFM @ 3 Ounces
Spirax Sarco	0.500	0.750	0.966
All of the following Spirax Sarco traps use the same air-vent module, so they will vent at about the same rate regardless of the trap's size: FT-15, FT-30, FT-75, FT-125, FT-150, FT-200, FT-20, FTB-20, FTB-30, FTB-125, FTB-175, FTI-15, FTI-30, FTI-75, FTI-125, FTI-200			
Hoffman	0.500	0.750	0.950
FTO15H, FTO15C, FTO15X, FTO151; 55 Series all use the same air-vent module regardless of trap size			
Stirling	1.300	2.083	2.666
All Stirling F&T traps (from vacuum to 15-psi) use the same air-vent module regardless of trap size			
Barnes & Jones	0.750	1.200	1.500
All Barnes & Jones F&T traps (from vacuum to 15-psi) use the same air-vent module regardless of trap size			

As you look at those numbers, pay close attention to what happens to a trap's venting rate when you rebuild it with a parts kit. Consider, for example, the Hoffman 17C, half-inch trap with the factory parts in place. At three ounces of pressure, the trap will pass 1.100 cubic feet of air. If you replaced the trap's original parts with a Barnes & Jones Cage Unit, the venting capacity of that trap would more than double (to 2.70 cubic feet per minute).

If you replace some, but not all, of the trap parts in an apartment-house riser, suddenly the radiators with the Barnes and Jones Cage Units will heat faster than the radiators with the original parts. This may be a good thing or a bad thing, depending on where the radiators are located.

There's plenty of food for thought in that chart.

An interesting aside

Gerry Gill and Steve Pajek were working on an old, two-pipe vapor system. This one had a steam main that fed all the risers to the two-pipe radiators, and ended by dropping into a wet return, which brought the condensate back to the boiler. The end of the main had a large air vent. The dry return picked up the condensate from the radiators, and also gave the air in the radiator risers, and the radiators themselves, a way to pass to a large main vent at the end of the dry return. To put the condensate back into the boiler, this job had a boiler return trap, which is essentially a mechanical condensate pump.

This drawing is very close to what Gerry and Steve were dealing with.

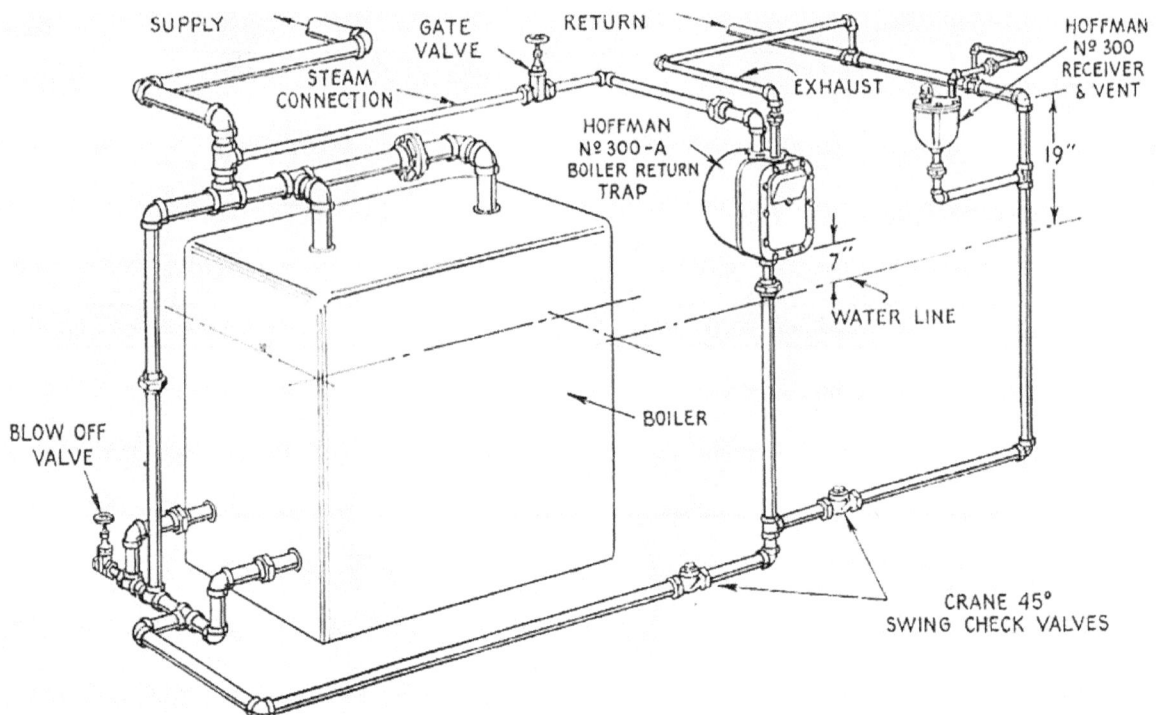

So follow the sequence. On start-up, air fills all the pipes and radiators above the boiler waterline. Steam forms and nudges the air from the mains, the risers and the radiators. The steam pushes the air in the main toward that big air vent near the end of the main. It also pushes the air in the risers and radiators toward that air vent at the end of the dry return.

Now think like steam. Suppose you had to travel all the way to the end of the main before you were allowed to climb any of those risers and enter the radiators. If you had to do that, the system would be in better balance, right? You'd enter all the risers at the same time.

The way to make that happen would be to open the air vent near the end of the steam main, *while keeping closed the air vent that's at the end of the dry return*. Where there is air steam will not go.

So this is what Gerry and Steve did. They piped a small, steam-rated zone valve on the pipe leading to the air vents at the end of the dry return. You can see it in this photo.

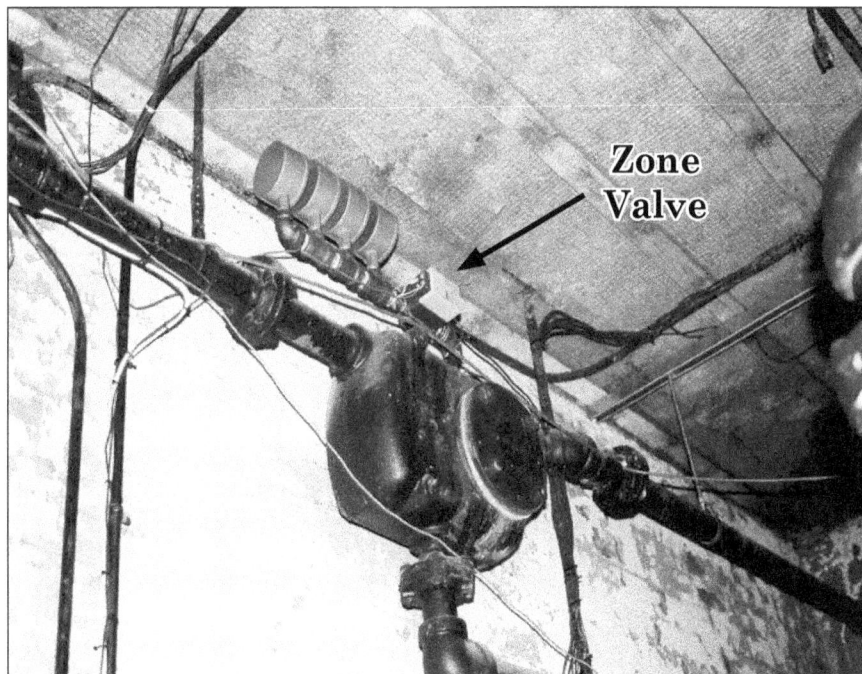

Air can't leave those vents until that zone valve opens, and none of the radiators will get hot until that happens. What opens the zone valve? That aquastat that you see in the photo on the following page.

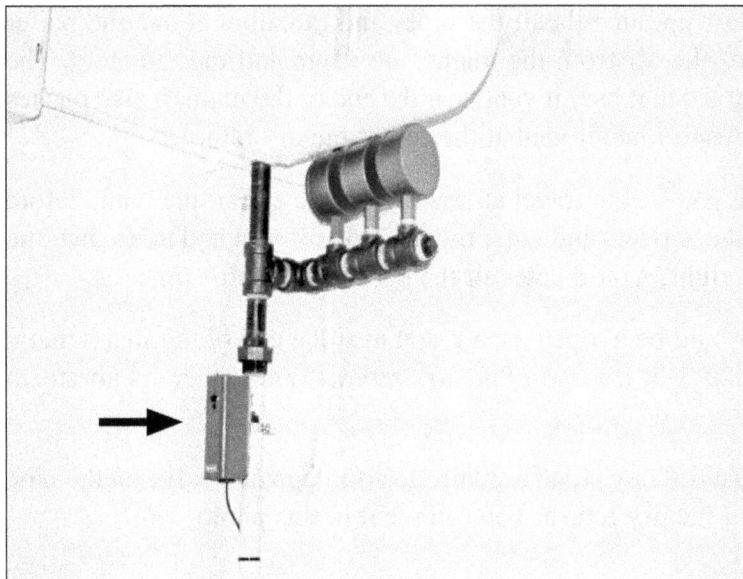

Aquastat at the end of the steam main opens the locked air vents at the end of the dry return.

That's the end of the steam main. Those air vents are wide open and ready to vent all the air that's in the main when the steam comes up. The steam will arrive at the aquastat, proving that the entire main is filled with steam and ready to flow into all the risers at the same time. The aquastat will feel the steam's heat and send a signal to the zone valve that it's now time to open. When the zone valve opens, air flows from all the radiators at the same time and steam enters evenly across the whole system.

Even steam distribution saves money. That's *greening* steam!

Steamhead

Frank Wilsey, known to his friends and customers as "Steamhead," works with Gordon Schweizer at All Steamed Up, Inc. (310 Centre Avenue, Towson, MD 21286. Phone: 410-321-8116). Together, these two guys do amazing things with the old steam systems of Baltimore and the surrounding area.

Steamhead told me about a job that he and Gordon had recently completed. It was a co-op apartment building, and I'll give him the floor:

"This building dates to 1915. It has 32 apartments, each with two or three bedrooms. The heating system is one-pipe steam, about 6,100 Square Feet EDR and there's a pumped return line.

"The current boilers are two, Slant/Fin Series 80, each rated at 7,285 Square Feet EDR. They have PowerFlame gas burners with on/off firing.

"The problem was outrageous fuel consumption, and poor heat distribution to the ends of the building. We added main vents to the places where they belonged. We used 25 Gorton #2 vents, one Gorton #1 vent, and two Hoffman #75 vents. We also rewired the control system so only one boiler could fire at a time (that's all they needed). We also replaced an ailing outdoor-air-temperature reset controller with a tekmar #269 and properly placed sensors. Then, we vented the risers with Gorton #D vents, which equalized steam flow between the

floors. We flushed out the boilers and tuned the burners with a digital combustion analyzer. The savings on their $35,000 annual fuel bill was 32 percent. Steam now fills the entire system instead of only one-third of it. The projected payback was three years.

"By the way, the building's resident historian told us that the steam system had never worked right since it was installed. The original building owner sued the installing contractor over it. Eighty years later, we come along and it now works. Can you imagine how much fuel – first coal, then oil, then gas – it wasted during all that time?"

A 32% savings in annual fuel usage, with just a bit of tweaking and vents that are the proper size and in the right place. That's greening steam, and all it takes is know-how, and the belief that since all of this is just a blend of mechanical things and simple physics, it *can* work. Steamhead and Gordon prove that every day.

And as Dizzy Dean once said, "It ain't bragging if you can do it."

A word about thermistors

In his comments, Steamhead mentions the outdoor-air-temperature controller and the properly placed sensors. This is worth a few more words because that sensor can make or break the system.

Many big steam-heating systems, particularly those in apartment buildings, use some sort of heat-timing device to start the burner. They basically work like this:

When it gets cold outside, the burner will start. Local law often dictates the exact outdoor temperature that will trigger the start of the cycle. For instance, let's say that your city wants the heat on in all the rental properties whenever the outdoor temperature drops below 50 degrees. The heat-timing device will have an outdoor sensor that feels for that 50-degree air.

But who places that sensor, and what's happened to it since the day that person placed it? What if that sensor is on the sunny, south side of the building? In that position, it can pick up enough radiant heat on bright days to keep the heat in the building off. Or what if the sensor is located right next to the discharge from the clothes dryers in the laundry room? The heated air from the dryer will keep the boiler off (yes, I've seen this happen).

The sensor belongs on the shady (preferably north) side of the building, away from any source of heat, other than what comes from the air, and high enough so that curious fingers will leave it alone.

Okay, so it's now 50-degrees outside and the burner is running. How long should it run on that particular day? This brings us to the thermistor. This is the control that tells the heat-timing device when steam has reached the end of the main. Theoretically, when steam is at the end of the main, it should also be at all the radiators. At that point, the boiler will run for a certain amount of time before shutting down. How long the boiler runs depends on the outdoor temperature at that time. For instance, if it's 40-degrees outside, the boiler may run for 20 minutes out of an hour. If it's 20-degrees outside, the boiler will run longer because it's colder outside. Simple. The thermistor is the device that *starts* that countdown cycle to burner-shutoff, based on the outdoor temperature, and *where* you place the thermistor will play a large part in how much fuel you use.

So when you're troubleshooting one of these jobs, you have to find that thermistor, which is easier said than done because it could be anywhere, and it's probably buried under pipe insulation. Most of the time, the thermistor is at the end of the longest steam main. The thinking is that when the main gets hot to that point, steam is probably everywhere else, which may or may not be true, depending on the condition of the air vents and the steam traps throughout the building.

But there are knuckleheads in this world and they never rest, which is why I've seen thermistors installed on wet returns, way below the point where steam will ever go. You can't sense steam temperature on a pipe that's always filled with condensate. The burner will run forever in this case, and that's not green.

I once saw a thermistor installed on a sewer pipe. Why? Because there are knuckleheads in this world.

I once saw a thermistor installed on a gas line.

How come? The knucklehead was trying to save gas.

And because these people go to work every day of the week, it pays for you to get out of the boiler room and wander around.

CHAPTER 7

GREENING THE TRAPS

A steam trap's obituary

When a steam trap dies, it usually does so in its open position, and it doesn't say anything to anyone in the building. It fails open because this is the position that will cause you the most trouble.

The trap's demise doesn't appear in the papers. It doesn't clutch its chest and scream. It doesn't make the nightly news. It just quietly dies, and almost always wide-open.

Steam now has access to places where it doesn't belong, and that's never a good thing. It gets into the returns, which are too small (by 1,700 times) to handle steam. It does this because high pressure goes to low pressure, always.

With steam in the returns, air now gets trapped between the two relatively high-pressure points – the steam in the supply, and the steam in the return. The air can't get out, and where there is air, steam will not go (I think I've told you that a few times already).

So, many of the rooms go cold, and the condensate that does form in the radiators that do get hot can't drain down the return risers because they're filled with the steam that's trying to go up.

And if the water can't get back to the boiler, you're going to have problems with the water level *in* the boiler. The water hangs around in the system, creating water hammer. The heat in the building is uneven and people are opening the windows. They're banging on the pipes. They're calling at all hours.

None of this is green, so let's look at those traps. They're *real* important.

Steam traps trap steam

That's a trap's main job. Its other (and first) job is to pass air through itself so that steam can reach the radiator. Remember what our friend Gerry Gill said, "Before a radiator trap is a radiator trap, it is an air vent." Failed traps keep air from venting.

You'll see steam traps on larger, two-pipe jobs because one-pipe steam has its limits, especially in big buildings where the condensate often flows counter to the steam. The traps separate the steam from the condensate and keep peace in the valley. They allow for smaller supply pipes, since these pipes don't have to handle returning condensate.

I've often thought of traps on a steam system as being similar to balancing valves on a hot water system. They keep the hot side hot and the cold side cold, and they establish the points of pressure and no-pressure. They allow for the flow of steam, and they last for about 10 years

They get changed about every 50 years. Maybe.

And if you're wondering why this is so, I'll give you a straight answer:

1. New traps cost money.

2. They don't all fail on the same day.

3. They don't scream when they die.

4. It's often difficult to get at them.

And because of this, many people have come to believe that it's okay to have failed steam traps. Trust me; it's *not* okay. And it's certainly not green.

How traps work
(and why you should keep them working)

Let's start with the large ones. In steam heating, these will mainly be Float & Thermostatic traps, which go at the ends of the steam mains and the base of steam risers. Their job is to first allow air to pass, then close to steam, and finally, to drain the condensate into the dry-return mains so that it can flow back to the boiler.

Again, a dry-return main is any return pipe that's positioned higher than the waterline of the boiler. You'll find F&T traps on systems that have gravity returns, as well as those systems that have pumped returns. These traps keep the steam from working its way backwards into the condensate-return piping that's serving the radiators.

When they're at the base of steam risers, they drain the condensate that forms when the riser first heats up. Usually, a ¾" F&T is all you need because you're not going to get much condensate from this pipe. If you oversize the F&T drip trap, it will wiredraw and fail long before a properly sized (and much smaller) trap will. That's a waste of green.

How an F&T trap works

An F&T trap is both normally open and normally closed. The thermostatic element, which is over there on the left in the drawing, is normally open so that air can move through the trap ahead of the steam, and vent from the system somewhere beyond the trap. When steam arrives, the thermostatic element senses the heat, expands, and stops venting. It also stops the steam from moving any further.

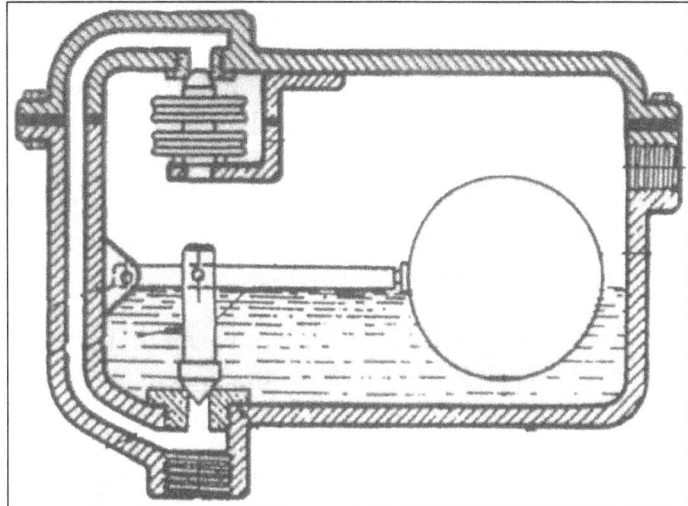

So now the steam is trapped and can't go anywhere. It's mixing with the condensate that's flowing into the trap and condensing. Now look at that round float that's connected to the rod that holds the pin that sits in the trap's seat. That part of the trap is normally closed. When enough water builds up inside the

Cutaway of F&T trap

trap, the float will rise with the tide and pull the pin away from the seat, allowing condensate to pass into the return. When all the condensate is gone, the trap will close again, and that's the full cycle. It happens over and over again.

Here's how to green F&T traps

1. Begin by sizing them properly. An oversized trap will fail because there's not enough water moving through it to fully lift that pin from the seat. The result is high-velocity water moving across the seat, which causes metal erosion (called wiredrawing). Within a heating system, if the F&T trap is line-sized, it's the wrong size. It's way too large and it will fail. This will *always* be true. Don't waste your green buying more trap than you need. It won't last.

2. Understand that F&T traps don't sense temperature. They're going to dump condensate at saturated-steam temperature. The higher you run the pressure, the more flash steam you'll wind up with in the return lines, and that can cause water hammer.

3. You can't check the operation of an F&T trap with a thermometer because they don't sense temperature. The best way to know whether or not the trap is working is to crack a union or open a valve just downstream of the trap. If live steam is

continuously passing, the trap needs attention. Be careful when you're doing this. Steam is hot.

4. If possible, set up a trap-testing station within your building and check the traps during the warmer months. That way, you'll be ready for the colder months. It pays to check them every year.

5. F&T traps, like all steam traps, work on differential pressure. The trap itself doesn't have the ability to move condensate and air. All it can do is open and close in response to the difference in pressure between its inlet and outlet. Control valves on heating equipment (such as unit heaters) will shut when the temperature is just right within the space being heated, and that takes away the inlet pressure to the trap. If you don't size the trap properly, water won't drain from it. If that trap is serving a coil that's heating, and it's not the right size trap, the coil can freeze and break when the control valve shuts. If you're not sure how to size traps, call one of the trap manufacturers and draw on the experience of their local rep. Those people are happy to help. I know; I used to be one of them.

6. F&T traps need strainers upstream to keep the debris out of them. Someone has to check and clean those strainers at least once a year.

How a thermostatic radiator trap works

These go on the radiators, at the outlet side, and they *do* sense temperature. They're normally open and air will pass through them ahead of the steam, and then vent from the system somewhere downstream.

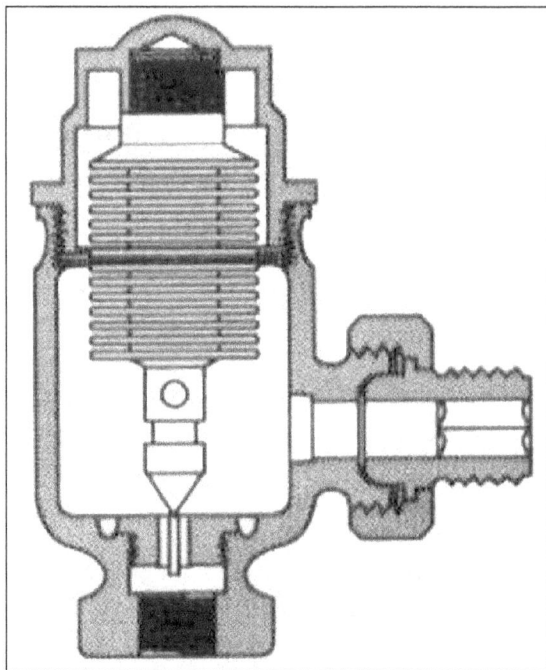

Cutaway of a thermostatic radiator trap

When steam arrives, the thermostatic element will expand because it's partially filled with alcohol and under a vacuum. Alcohol boils at a temperature that's lower than the temperature at which water boils. When the element expands, it pushes the pin into the seat and traps the steam inside the radiator.

Condensate builds up and flows into the trap. At this point, we need a drop in temperature of about 10-15° Fahrenheit before the trap will reopen. When it does open, the differential pressure across the trap will move the condensate from the radiator into the return so it can make its way back to the boiler.

Because they're sensitive to temperature, you can check a thermostatic radiator trap with a thermometer. Look for that 10-15° difference in temperature, and keep in mind that this temperature is relative to the steam pressure. Steam at 1-psi pressure is about 216°F, so on a low-pressure, space-heating system, the traps will open when the condensate is somewhere between 200° and 205° F. If you're running, say, 10-psi pressure on that space-heating system because you happen to be a knucklehead, the steam temperature will be 240° F, and the trap will vomit condensate at about 225°F into the return, where some of it will flash right back into steam, partially defeating the purpose of the trap.

Trap testing gone to pot

If you'd like to have more fun than a thermometer can provide, you can make yourself a trap-testing station out of a cooking pot, as this adventurous fellow did.

Thermostatic radiator traps will open and close hundreds of thousands of times each heating season, and manufacturers give them about a 10-year lifespan. As I said earlier, they get checked about every 50 years, and that's one of the reasons why steam heating is noisy and inefficient. Would your car run efficiently if you never changed the oil or tuned the engine? Neither would mine.

The problem with checking traps, though, is that we sometimes can't get at them. And the wealthier the tenants, the worse the problem becomes. I've been in some Manhattan apartment buildings where the rich folks thought it was a fine idea to encase the radiators in marble, leaving no access to the steam traps. The marble looked marvelous, I have to admit, but the air couldn't get to the radiators to warm the rich folks (which is why I was there), and the traps were entombed better than Ulysses S. Grant.

I looked at the rich people.

They looked at me.

I smiled.

Winter work?

Around 1975, when I was working for the rep in New York City, we sold hundreds of thermostatic radiator valves to the people who ran a cooperative apartment building. We also sold them an equal amount of new thermostatic radiator traps. This was right after the first OPEC oil embargo and everyone in NYC was abuzz with talk of saving fuel. A combination of new radiator traps and TRVs seemed like just the ticket.

Now, when do you suppose they decided to do this work? Do you think they did it during the summer when the steam heat was off? Nah, that's a silly time to work on heating. It's too warm. Summer is the time to work on air conditioning. Everyone knows that.

So they were adding these TRVs and new steam traps to lines that were under steam pressure. They'd isolate one line at a time, of course, so as to not get burned, but they couldn't get into all the apartments on that line on the same day, and they didn't want the tenants to complain, so they turned the line off during the day, and opened it again at night.

Now, think of that steam line that rises, say, through the living room of all the stacked apartments from the first floor to the 20th floor. Let's install new TRVs and radiator traps on the first three floors today. That's about all we can get done because we're using our own building people to do the work.

The last time anyone worked on these traps, F.D.R. was sitting in the Oval Office. There's been steam in the return lines, raising both hell and the fuel bills since the J.F.K. years. So the building staff installs the new TRVs and traps in those three apartments and turns on the steam that night. The TRVs close on temperature, and so do the new traps. The steam condenses inside the radiators with the new equipment and forms a vacuum because the radiator is now closed on the inlet side (thanks to the TRV) and on the outlet side (because of the closed trap).

With me so far? Okay, keep in mind that traps work on pressure differential, and that high pressure goes to low pressure – always. When the new steam traps opened, the steam in the return line was waiting. It rushed backward into the radiators with the new traps and the resulting water hammer damaged the new trap elements. It also overheated the radiators with the closed TRVs. Nature hates a vacuum.

I spent a lot of hours on that job with the factory guy from the TRV company. We both got a good education. The main thing we learned was that *summer* is the time to work on TRVs and traps.

Oh, and before we leave thermostatic traps, please go back and read again the part of the chapter on Greening the Venting where Gerry Gill and Steve Pajek's chart shows you the difference in venting rates across a new thermostatic trap, fresh from the manufacturer, and one that contains a repair kit.

An unbalanced steam system will waste green.

One trap for the whole building?

When steam traps go bad, they pass steam into the return lines, which are too small to handle steam, so you get water hammer and unbalanced heat distribution to the radiators. The fuel bills and the maintenance bills soar and the tenants are miserable. It's not normal and it's not nice.

If the system has a condensate- or boiler-feed pump to put the condensate back into the boiler, the misdirected steam will spew from the vent on the pump's receiver and raise the humidity level in the boiler room to the point where it feels like a rain forest in there. High humidity within a confined space isn't good for anything mechanical or electrical. Oh, and it also gets very hot in there. Not nice.

The receiver has a vent because it's not rated to withstand any pressure, other than the slight static weight of the returning condensate. Often, a well-meaning building superintendent or a service person (who should know better) will plug that vent, which sets up a very dangerous situation because condensate- and boiler-feed pump receivers can (and often do) explode when someone plugs the receiver vent. That's the worst-case scenario. The not-as-deadly scenario is that the air now has no way out of the system and the steam stops moving (where there is air, steam will not go). Keep in mind that the vent on the receiver is usually the only vent on these two-pipe, pumped-return systems.

So here's what happens in real life. Rather than plug the vent, some well-meaning (but knuckleheaded) serviceperson will decide to place one huge steam trap right at the inlet to the pump's receiver. His reasoning is that he can stop the steam right there at the trunk of the tree. This guy never stops to ask that crucial Dead Man Question: If that could be done, why didn't the Dead Man do it?

Seriously. If the Dead Man could have gotten by with one huge steam trap at the receiver rather install hundreds of steam traps at the individual two-pipe radiators, the ends of the steam mains, the base of the supply risers, why didn't he do it?

Was the Dead Man that stupid?

Didn't he consider all the money he could have saved back in the day?

Ask that question and you'll see the folly of trying to get by with just one big steam trap. Steam traps trap steam, and if you let steam run wild through the radiators and enter the returns, bad things will happen to that system. Sure, using that one trap just before the condensate- or boiler-feed pump's receiver may stop the steam from spewing into the boiler room, but at what cost? And think about what happens when that one big trap finally opens. Since the steam distribution has gone to pot, it's likely that someone has cranked up the steam pressure. That big F&T trap at the "trunk of the tree" is now going to dump red-hot condensate into the receiver, and much of that hot condensate will flash right back into steam and wind up in the boiler room. And the hot condensate is liable to cause the pump to cavitate and fail long before its time, presenting you with even more expense.

If it could be done, the Dead Man would have done it.

Never try to get by with just one trap for the whole building. It doesn't work and it leads to balance problems, water hammer and higher fuel bills

The ladder

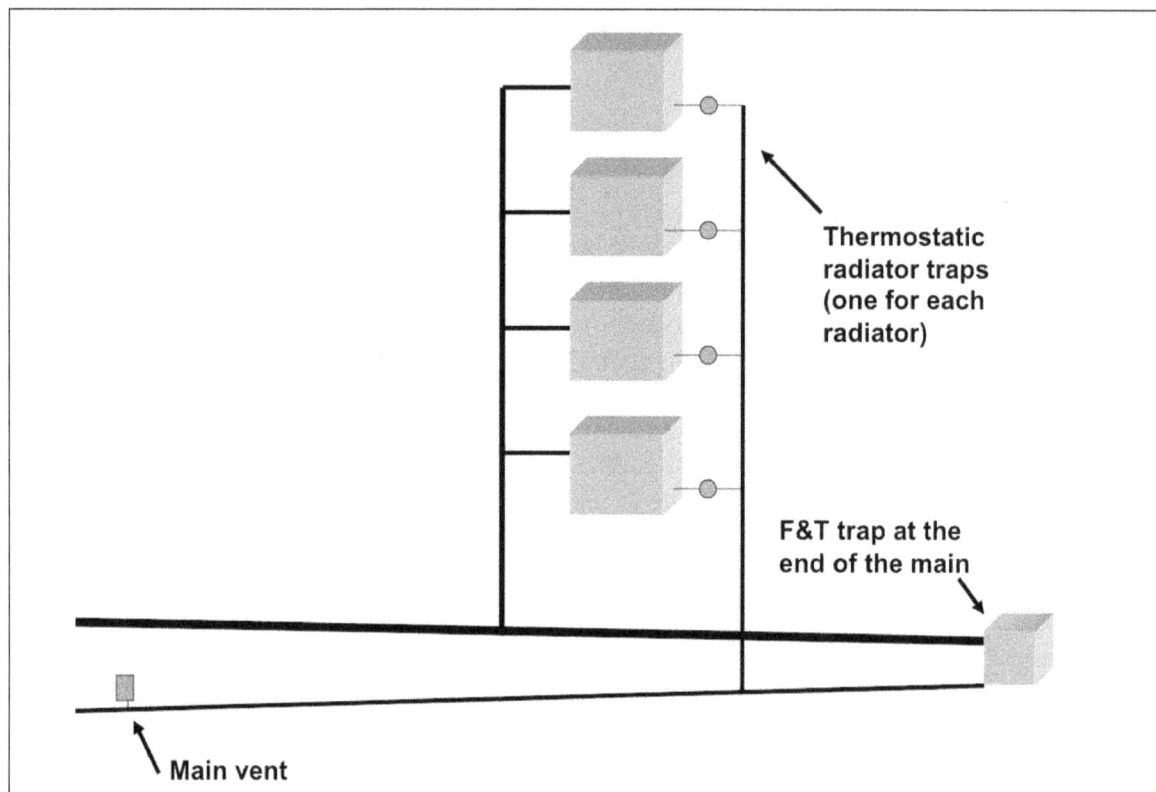

Two-pipe riser

Here's why fixing radiator traps saves green. Take a look at that sketch and notice how much it looks like a ladder. The steam supply riser on the left is one side of the ladder; the return riser is the other side, and the radiators are the rungs. Notice that there's a thermostatic radiator trap at the outlet of each radiator. There's also an F&T trap at the end of the steam main. Its job is to keep steam out of the dry return main so that there's a pressure differential across the radiators. Without a difference in pressure, nothing moves.

Now I'd like you to meet a couple of the tenants who live in this building. Joe lives on the first floor; Moe lives on the top floor. The steam climbs the ladder, pushing air ahead of itself. Joe is the first tenant to get heat; Moe is the last to get heat. For years, the steam traps on their radiators have worked. Joe and Moe have been happy men. The traps keep the hot side hot and the cold side cold and the pressure differential keeps the air moving from the radiators so the steam can flow freely and evenly.

But then one day, the trap on Joe's radiator fails. Its element gets tired and breaks, leaving Joe's trap in the open position. Joe's failed trap is about to screw up the heat in all the apartments above him, and especially in Moe's apartment because Moe's on the top floor, but here's the delicious part:

Joe will still have heat.

Joe will *never* call you for help.

He's messing up everyone else and he doesn't even know it. Nor does he care. You'll show up at the building to troubleshoot the problem, but you won't go to Joe's apartment because Joe isn't complaining.

You'll go to Moe's apartment because Moe is miserable.

You check Moe's radiator and confirm that it's cold. You decide that there's something wrong with the steam trap, so you change that trap's guts. It makes no difference. Moe is still cold and miserable.

You change the parts again. It makes no difference. You try again. No difference. You blame the parts manufacturer for a bad run of trap guts.

Moe has no heat because the air can't get out of his radiators. There's steam in the return riser because of Joe's failed trap. Where there is air, steam will not go, and the condensate that does form finds it difficult to flow back to the boiler because of the pressurized return. Oh, and steam is also flowing in the dry return now because Joe's failed trap is giving it a shortcut around that end-of-main F&T trap.

The boiler goes off on low water because the condensate isn't returning quickly enough. You install an automatic water feeder. The boiler floods.

Or if this system has a condensate receiver, it overflows.

You blame the manufacturer.

Frustrated, you decide to add a one-pipe steam air vent to Moe's radiator. This allows the air to escape and gives Moe the heat he's been craving. It works because of the First Law of Steam Heating, which, once again, states, *When you do something stupid, you will always get a reward, which leads you to do things of even greater stupidity.*

But have you given it the Dead Men Test? If one-pipe radiator vents worked on two-pipe radiators, why didn't the Dead Man install them?

Why?

And the answer is the Dead Man didn't install them because, while those vents allow the air to escape, they also mask the fact that the traps have failed. Steam is in the returns; it's keeping the condensate from flowing freely back to the boiler. It's causing water hammer that's damaging the pipes and depriving the tenants of sleep. It's causing the boiler to flood. It's making people open their windows on frigid days. It's causing people to move out of the building. And things will only get worse because now that Moe has heat, you'll probably add air vents to all the other radiators as well.

Please don't do this. It wastes green.

A word about traps on gravity-return systems

I've seen many apartment buildings with two-pipe, gravity return systems. They look like this.

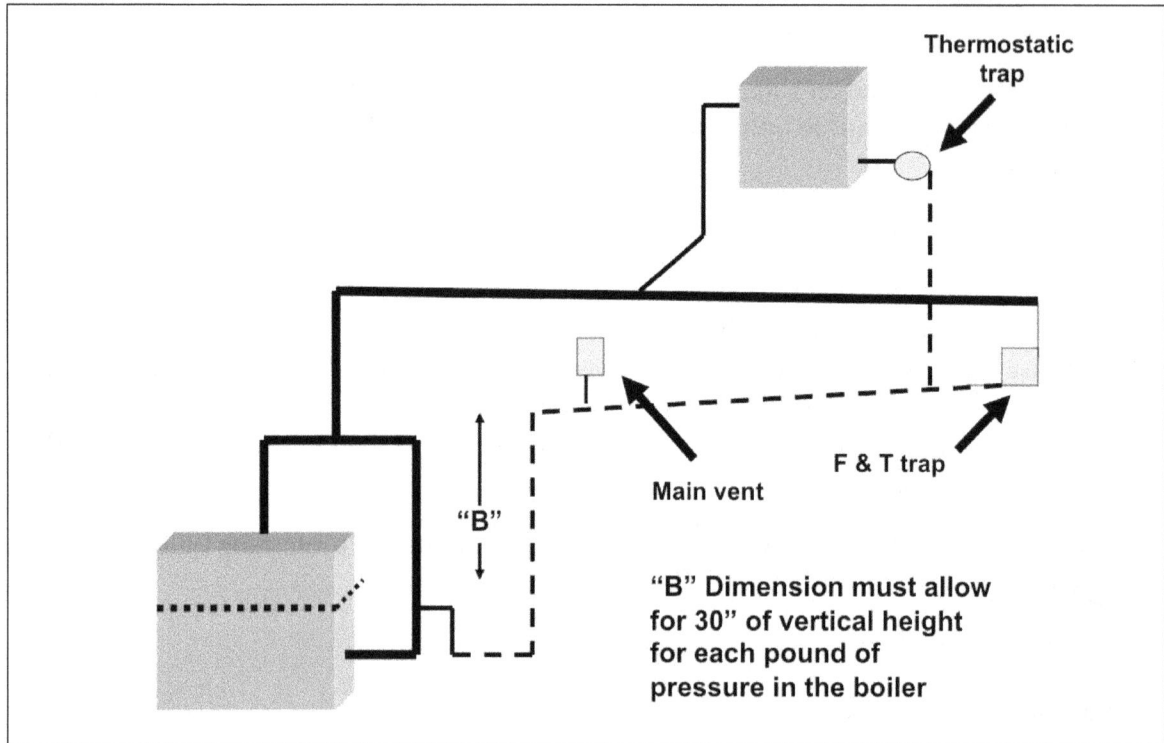

Two-pipe, gravity return system

It's a nice design for an apartment building because, without the condensate- or boiler-feed pump, there's less equipment to maintain. But having this type of system means you have to keep the boiler pressure low or the water in the boiler will back into the return and block that main air vent.

Look at the drawing and note the dimension I have marked as "B" because that vertical distance determines the boiler's maximum pressure. The steam traps on the radiators and at the end of the steam main keep steam pressure out of the return lines. That means that the only pressure we have available to overcome the pressure that's inside the boiler is the static weight of the water in the "B" dimension. A column of water 28" high exerts one pound per square inch of pressure at its base, so for each pound of pressure in the boiler you'll need at least 28" of "B" Dimension. I like to allow a bit more height to overcome the friction in the wet return below the "B" Dimension, so let's allow 30" of height per psi of boiler pressure.

Look again at the drawing. Let's say there's 45" of height in the "B" Dimension. That's equal to a static pressure of just over 1-1/2-psi. As long as the boiler operates at that pressure, or lower, the water will return to the boiler and air will be able to leave the system through that main vent. But suppose someone comes along and cranks the boiler pressure up to 3-psi. Now you need a "B" dimension that's equal in height to 90". Without it, water will back into the dry return and block the main vent. At that point, the air won't be able to escape from the system because that main vent is the *only* vent on this system.

You'll usually see the problem at the top floor radiators because the steam is climbing the "ladder" and the air has already escaped from the radiators on the lower floors.

Lower the boiler pressure and the water will recede, allowing the main vent to do its job. You'll look brilliant.

When in doubt, crank it down.

CHAPTER 8

GREENING THE ENVELOPE

This is going to be a *very* short chapter because I think most of this is common sense.

In any building, we put in the heat, and the heat then leaks out. The tighter the building envelope is, the less inclined heat will be to leak out. It's that simple.

As I write this, people are hard at work on the Empire State Building. They're bringing that landmark building into the 21st Century by replacing 6,500 old windows with new, insulated windows. They're adding reflective insulation behind all of the radiators. They're installing smarter controls and they're upgrading the chillers. They expect to invest $20 million in this major renovation, but they also expect to save more than $4 million each year on fuel, which will give them a five-year payback.

And that's as green as it gets.

Insulate. Insulate. Insulate.

CHAPTER 9

GREENING YOUR OUTLOOK

So let me ask again that question we asked ourselves right at the start of this book: Why the heck, in the 21st-Century, would anyone want to heat a building with 19th-Century technology?

Let's face it, there are heating systems available these days that are better than steam. Radiant heating systems, for instance, are more efficient because they operate at low temperature and heat people and objects without heating the air. Low-temperature hydronic systems that use panel radiators and condensing boilers are also very efficient and use less fuel than the typical steam system will use (providing they're in a well-built, tight building).

Ground-source heat pumps can extract the natural heat from the sun, which the earth stores below its surface, and heat buildings very efficiently. These systems, however, are costly to install because you have to bury a lot of pipe in the ground to get at that natural heat.

Geothermal systems pull hot water from deep below the surface and they're nice to have if your building is above such a source of hot water, and if you have the money to invest in the drilling.

The ultimate in heating is the Passive House, which involves no heating system at all (humans come with their own heating systems). The walls in a Passive House are very thick and well-insulated, and the windows are remarkably efficient. Passive Houses have heat-recovery ventilators to keep the natural heat from people, as well as the heat from any appliances, inside the building.

These technologies are modern and viable and very interesting. I love them all, but you should consider the payback period before making a decision. And if you do have an old steam-heating system, think about using what you've learned about tweaking. You can save a lot of green by tweaking.

I've found that a lack of knowledge scares a lot of people. You've taken the time to read this book, and you may have read other books about steam heating, so you've gained the knowledge. You know what the Dead Men knew. I hope you now see that none of this is that complicated. Those who don't take the time to learn will *always* think that it's complicated, and they'll shy away from it, and try to convince everyone around them to tear out those old systems and begin anew.

But that may not be practical. And it may not make economic sense.

Steam is a magnificent way to move lots of Btus from one place to another. There's really nothing better when it comes to flat-out moving heat, but steam is also an old way of heating.

Does that mean that it has to go?

That's really for you to decide.

Will there eventually be laws banning steam in the U.S.? This happened in many European countries. The governments passed laws stating that the maximum temperature leaving any heating boiler could be only so high, and that's how they got rid of the steam. But keep in mind that these were countries that were rebuilding after two world wars. Many of their buildings were in ruins and they were starting from scratch.

Do you see a time when our politicians will outlaw steam heating? They seem to have more-pressing issues to deal with right now. It's probable that, in time, we'll shift away from steam as many parts of America have done, but consider a city such as New York that's filled with tall, steam-heated buildings.

Could you get all of those co-op boards to agree, unless there was a law forcing them to change their heating systems? I've yet to meet a co-op board that is eager to spend money, especially on a heating system. Usually, they arrive at this decision kicking and screaming, and almost always in response to some emergency.

No, unless there's a law banning it, I think steam heating will be around for some years to come. In the meantime, go after that low-hanging fruit. You can do a lot with a few bucks. Go back to the chapter on Greening the Venting and reread what Steamhead did in that apartment building. A small investment resulted in a 32% savings in fuel. And there are so many other examples of this. You don't have to spend a fortune to tweak a steam-heating system, and the results are impressive.

And think of all our historically significant buildings. What if we could get them working more efficiently without changing the entire system? Look at the Empire State Building as an example. They'll be enjoying huge savings in their fuel bills, and a five-year payback. Why tear out the steam and begin anew? It makes no sense.

I want you to be able to do what others say can't be done, and now you're well on your way to doing that. Working on an old steam system is a lot like working on a classic car. You can rebuild the engine (the boiler). You can work on the body (the piping network). You can fix the upholstery (the radiators). Why junk a classic when, with a bit of thought, planning and work, you can make it gleam?

Besides, being able to do this brings a certain sense of pride. And, if you're a heating professional, it also brings new business. When you can do what others say is impossible, people will find you. Trust me.

It comes down to how much you know, and how willing you are to learn more, and to tweaking. It *is* possible to green steam, but you first have to *believe* that you can. I hope you now believe that, and that you'll get out there and do the things that others say can't be done. And if you have questions as you go along, bring them to HeatingHelp.com. We love good questions, and we never close.

Thanks for listening. Go get 'em!

INDEX

M

Main vents
    Capacity, 131-132
    Location, 124
Master trap, 116-117
Master venting, 125-128
MEX, 80, 103
Mills, John Henry, 13-15
Mills Rule (2-20-200 Rule), 13-14, 38
Mills System, 13
Mud (in boiler), 73-74
Murdough, Noel, 129
Murray Hill Hotel, 55

N

New York Cares, 129, 139
New York City's Department of Housing
    Preservation and Development, 49
New York Public Library, 2
Noise, 10
Northern Indiana Brass Company, 2

O

Oil (in water), 70-73
OPEC Oil Embargo, 77, 84, 153
Orifices, 62, 111-112

P

Pajek, Nate, 64
Pajek, Steve, 62-66, 128, 140, 144-146,
    154
pH, 75-76
Passive House, 163
Pigtail, 104-105
Pocket Rocket, 130
PowerFlame, 146
Pressuretrol
    Location, 104
    Settings, 53-56, 59-60

Piping Pick-up Factor, 19-20
Pipe size, 47
Priming, 74

R

Radiant heat, 163
Radiators
    Column, 23, 24
    Converting to hot water, 23-24
    Enclosures, 33-34
    First, 16
    Indirect, 37-39
    One-pipe steam, 20-21
    Painting
        How-to, 35-36
        with metallic paint, 29-32
    Paint color, 32
    Sizing, 16, 23-26, 28-29
    Thin Tube, 25-26
    Two-pipe, air-vent, 22, 44-46
    Two-pipe steam, 21-22
Radiator vent capacity, 135-139
Reed, J.R., 15
Reflective Insulation, 26
Return piping, 113
Return valve, 102-103
Risers
    From boiler, 85
    To radiators, 109-110

S

Schweizer, Gordon, 7, 146-147
Skimming (boilers), 71-73
Slant/Fin, 146
Square Feed of Radiation, 16
Spanish Influenza, 28
Steam traps, 47, 149-159
    Venting capacity of, 140-143
Steam velocity, 50-52
Strock, Clifford, 126
Supply piping, 107-108

Supply valve, 111
Surging, 74

T

tekmar, 146
Thermistor, 147-148
The Ideal Fitter 4
The Lovely Marianne, 5
The Lost Art of Steam Heating, 5-6
*The Principles of Warming and Ventilating
    Public Buildings,* 57
The Steam Book, 5
Thermostatic radiator trap, 152-153
Trane, 112
Trap tester, 153
Tredgold, Thomas, 57-58

U

United States Board of Health, 28-29
U.S. Department of Commerce National
    Bureau of Standards, 29-32

V

Vacuum systems, 56-57
Valves
    Pneumatic, 41
    Thermostatic radiator valve for one-pipe
    steam, 43-44
    Thermostatic radiator valve for two-pipe
    steam, 41-42, 153
Vaporstat, 61
Ventrite, 132

W

Water hammer, 10, 116
Water quality, 47
Water treatment, 79
Watt, James, 16

Weil-McLain, 72, 87-93
Wilsey, Frank "Steamhead", 7, 146-147, 164
World War II, 2

Z

Zone valves, 115